D0389185

DRINK THE BITTER ROOT

ALSO BY GARY GEDDES

. . . . .

POETRY

War & Other Measures

The Terracotta Army

Girl by the Water

Active Trading:
Selected Poems 1970–1995

Flying Blind

Skaldance

Falsework

Swimming Ginger

FICTION

The Unsettling of the West

NON-FICTION

Letters from Managua:
Meditations on Politics & Art

Sailing Home:
A Journey through Time, Place & Memory

Kingdom of Ten Thousand Things:
An Impossible Journey from Kabul to Chiapas

GARY GEDDES

DRINK THE BITTER ROOT

a search
for justice
and healing
in africa

COUNTERPOINT
BERKELEY

First published by Douglas & McIntyre: An imprint of D&M Publishers Inc.
2323 Quebec Street, Suite 201, Vancouver, BC V5T 4S7

Library of Congress Cataloging-in-Publication Data is available.

ISBN: 978-1-58243-788-0

Jacket and text design by Naomi MacDougall

Jacket photographs © Jo Harpley/Getty Images (top); © iStockphoto (bottom)

Map by Eric Leinberger

Printed in the United States of America

COUNTERPOINT

1919 Fifth Street

Berkeley, CA 94710

www.counterpointpress.com

Distributed by Publishers Group West

10 9 8 7 6 5 4 3 2 1

Contents

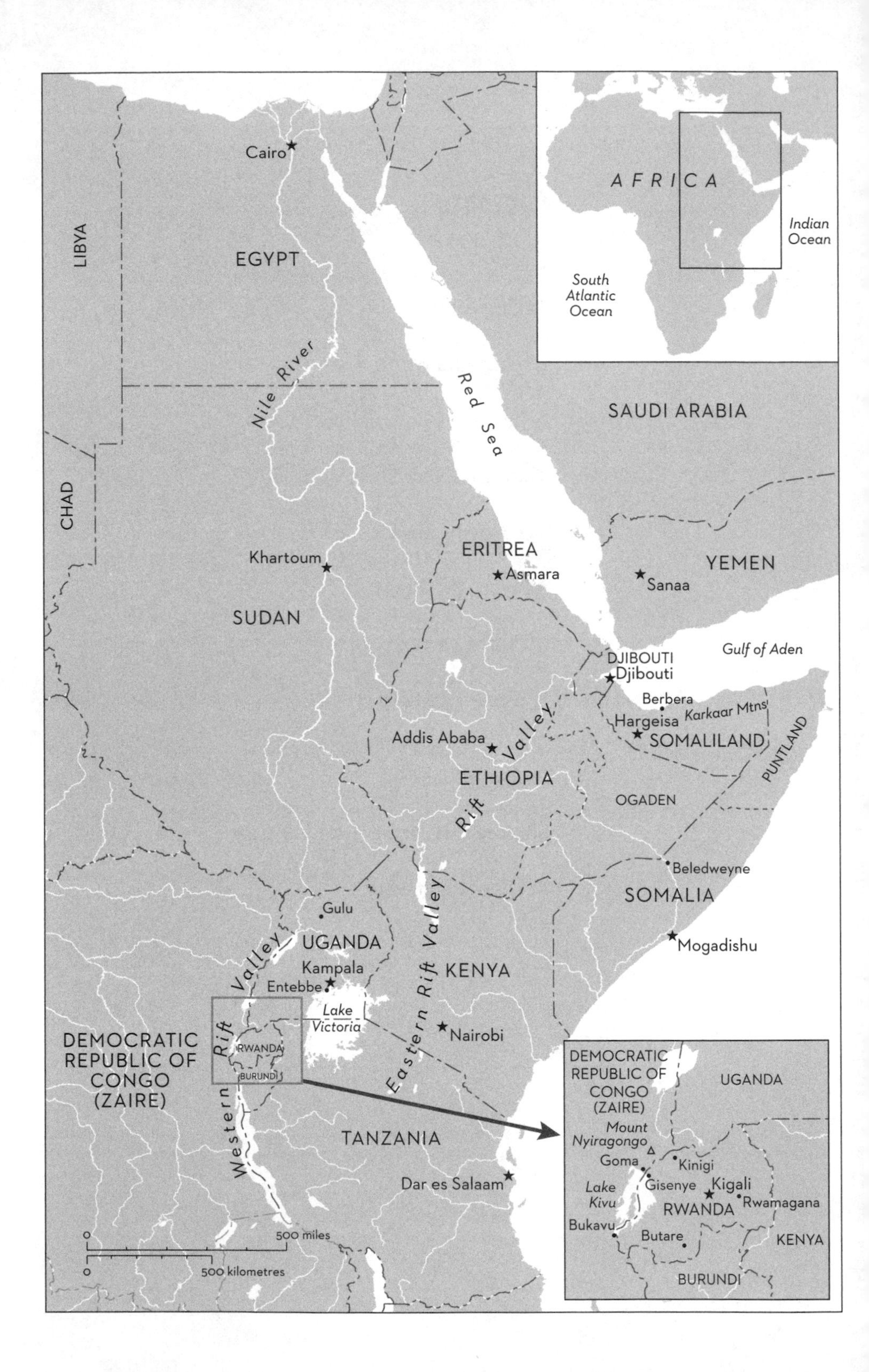

Cairo
LIBYA
EGYPT
Nile River
Red Sea
SAUDI ARABIA
AFRICA
Indian Ocean
South Atlantic Ocean
CHAD
Khartoum
SUDAN
ERITREA
Asmara
YEMEN
Sanaa
Gulf of Aden
DJIBOUTI
Djibouti
Berbera
Karkaar Mtns
Hargeisa
SOMALILAND
PUNTLAND
Addis Ababa
Rift Valley
ETHIOPIA
OGADEN
Beledweyne
SOMALIA
Mogadishu
Gulu
UGANDA
Kampala
Entebbe
Eastern Rift Valley
KENYA
Lake Victoria
Nairobi
Western Rift Valley
DEMOCRATIC REPUBLIC OF CONGO (ZAIRE)
RWANDA
BURUNDI
TANZANIA
Dar es Salaam
0
500 miles
0
500 kilometres
DEMOCRATIC REPUBLIC OF CONGO (ZAIRE)
UGANDA
Mount Nyiragongo
Goma
Kinigi
Gisenye
Kigali
Lake Kivu
RWANDA
Rwamagana
Bukavu
Butare
KENYA
BURUNDI

Somewhere our belonging particles
Believe in us. If we could only find them.

W.S. GRAHAM

Acronyms

AYINET	African Youth Initiative Network
CIDA	Canadian International Development Agency
CNDP	National Congress for the Defense of the People
CRONGD	Regional Council of Development NGOs
CTO	Centre de Transit et Orientation
DRC	Democratic of Congo
EPRDM	Eritrean People's Revolutionary Democratic Movement
ICC	International Criminal Court
OTP	Office of the Prosecutor
ICTR	International Criminal Tribunal for Rwanda
ICTY	International Criminal Tribunal for the former Yugoslavia
LRA	Lord's Resistance Army
MONUC	United Nations Organization Mission in the Democratic of Congo
NATO	North Atlantic Treaty Organization
NGO	Non-Governmental Organization
NRA	National Resistance Army
PARECO	Coalition of the Congolese Patriotic Resistance
SCF	Save the Children France
SHURANET	Somaliland Human Rights Network
TASO	The AIDS Support Organization
UPDF	Uganda People's Defense Force
U.N.	United Nations
UNICEF	United Nation's Children's Fund
UPDF	Uganda People's Defense Force
USAID	US Agency for International Development
WE-ACTX	Women's Equity in Access to Care and Treatment

one *Taking Aim*

IF QUESTION marks hang over a number of my undertakings, the one suspended over this journey was larger than usual. I imagined my wife and daughters looking up at the quizzical symbol—I pictured it as a pulsing red neon light in the shape of an inverted umbrella handle—and shaking their heads sadly. I shared their quandary. I could hardly breathe in the dust and intense heat, and my vertebrae were rapidly being pulverized from the impact of jagged rocks and potholes. I grabbed my hat and held on as the motorcycle bounced over the treacherous crust of lava and skidded around a corner, sending out a spray of debris into what had once been the main street of Goma, Congo's easternmost provincial capital, part of a region where warring militias had contributed to the deaths of four million people and the displacement of even more in the last fifteen years. Several armoured UN vehicles, with troops in full battle gear, roared past us on the left, heading in the opposite direction, away from the front lines. I saw only devastation, a blasted wasteland of ruined buildings. *Eruption* was the operative word here. Choose your enemy, Goma seemed to say: humankind or nature, it's carnage either way. We skirted the remains of a business with gaping windows. A swollen tongue of lava protruded from the doorway, making its implicit comment on the notion of "progress."

A kid wearing a blue T-shirt with the word "Westmount" across the chest shouted as we careened past: "Hey, *mzungu*, watch your ass!"

Good advice for a white man, and a warning I would think of often as I travelled through sub-Saharan Africa. It was not so much the statistical enormities that frightened me—800,000 butchered in a hundred days, for example—though these were terrible enough; nor abstractions such as greed, violence and corruption, which can be found anywhere. No, it was the personal stories, some shouted, some delivered in the faintest of whispers, others dragged screaming from the vault of memory. Alice, immobile, her heart turned to stone, describing, *sotto voce*, the atrocity she was forced to commit; and Nancy, sunlight reflecting off her mutilated face, highlighting the paler scar tissue, talking about forgiveness while I, presumably on her behalf, raged inside and plotted a slow, painful revenge on the perpetrators.

The reason I was in Africa had much to do with a notorious incident in Somalia fifteen years earlier, when a teenager named Shidane Abukar Arone slipped into the military compound where a regiment of soldiers from my country, Canada, were sweating out another African night on the perimeter of Beledweyne, a few hours' drive northwest of Mogadishu. Someone had left the gate open. The officers and men of Operation Deliverance—ostensibly a peacekeeping mission whose aim was to put a lid on the widespread violence and looting, spinoffs of the civil war still raging in other parts of the country—had bedded down an hour earlier. At least that's how it must have appeared to the young Somali, his face glistening in the moonlight and his hand, the one clutching prayer beads, unable to stop shaking. He saw bulk food containers, regulation gear, electrical equipment and spare parts for Jeeps and armoured vehicles. How much could he carry, especially if he were forced to run? Would it be food for his family or some item he could sell in the market? Shidane Arone was spared that decision, as his progress was being carefully monitored by one of the sentries on duty. Instead, he would discover the real meaning of deliverance, Canadian-style: hours of brutal torture and mutilation with cigarettes, metal pipes, boots, whatever his captors found readily at hand, his final utterance barely audible through the broken teeth and blood bubbling from his mouth.

The Somalia Affair, as it came to be called, is where my story wants to begin, but it is certainly not the beginning. The magnetic attraction of Africa started in my early teens with the example of medical missionary Albert Schweitzer, whose work amongst the Ogowe people in French Equatorial Africa convinced him that they were more deeply moral than Europeans. The life of Schweitzer, who had a thirst for knowledge and left behind promising careers in Europe as a writer, organist and theologian to study medicine and devote his life to this small community in the colonies, provided the stuff of dreams. Although I later abandoned my faith, or it abandoned me, the missionary zeal to do something of value never quite disappeared. So in 1964, while completing a diploma in education at Reading University in England, I applied for a teaching position at a Roman Catholic high school in Accra, Ghana. I was twenty-four. Shortly after the job was offered to me, reports started to appear in the international press that Ghana's president Kwame Nkrumah, who began his political career as an idealist, a socialist and a pan-Africanist, had turned against the unions and was jailing dissenters, including a number of foreign teachers. I turned down the offer, not realizing at the time that much of what I read was Cold War propaganda and that Belgium, the CIA and various neocolonialist elements were plotting to overthrow Nkrumah's left-leaning regime. But the sense of Africa as unfinished business, or a missed opportunity, remained.

In the decades that followed, I studied the works of Joseph Conrad, who had described the Belgian Congo as "the vilest scramble for loot that has ever disfigured the history of human consciousness and geographical exploration," and whose classic novel *Heart of Darkness* brilliantly dramatizes the brutalities of this scramble. I also read Chinua Achebe, Wole Soyinka and other African writers. Like any conscious observer of world affairs, I absorbed the plethora of disturbing media images of the so-called Dark Continent—gaunt faces of starving Biafrans, the Soweto uprising and bloodbath, the bloated bellies of children during the Ethiopian famine, the flight of refugees, the bodies of ethnic Tutsis hacked to pieces in the Rwandan genocide—and, more recently, a ceaseless litany of rapes, mutilations and murders taking place, with increasing impunity, in too many countries to name.

I felt both troubled and diminished by these realities and ashamed of my own privilege and inertia. I was equally uncomfortable with my ambivalent feelings about Africa, the only continent, other than Antarctica, I had not visited in my extensive literary travels. I'd written poems, articles and books about the Vietnam War, the Chilean coup and dictatorship, death squads in Guatemala, the U.S./Contra war against Nicaragua and the Israeli-Palestinian conflict. However, events unfolding in sub-Saharan Africa remained an enigma and a challenge. I felt a growing urgency to understand first-hand what was taking place and whether aid and intervention were making a difference in the lives of ordinary people. And I needed to hear answers to those questions from Africans themselves.

At sixty-eight, I had the freedom to consider such a journey. I still felt as passionate about these matters as I had in my thirties, perhaps more so. I appreciated the sentiment of Jean Vanier, founder of L'Arche communities for the mentally and physically disadvantaged, who had written: "Now, I'm free to do what I like, and what I like is to announce the message: that people who are weak have something to bring to us, that they are important people and it's important to listen to them. In some mysterious way, they change us. Being in a world of the strong, the powerful, you collect attitudes of power and hardness and invulnerability . . . It is vulnerability that brings us together."

With its fifty-four countries and infinite variety, Africa defies generalization. If Somalia was to be my final destination, I needed to narrow the search, make it somehow manageable. I was tempted to begin in Bas-Congo on the Atlantic coast, the epicentre of the slave trade, using *Heart of Darkness* as a shadow narrative, following Marlow up the Congo River to Kinshasa, then heading east through sub-Saharan Africa to Mogadishu. This narrow cross-section of the continent, which includes the Democratic Republic of Congo (DRC), Rwanda, Uganda,

Sudan, Ethiopia and Somalia, had become its most troubled and volatile region. However, although I understood that the moral darkness Conrad so brilliantly embodies is restricted to no one group, nation or geographical place but is rooted in the human heart, his famous novel has been politicized by African and post-colonial scholars and dismissed as racist. I did not want to be drawn into that debate.

More to the point, news reports and logistics—with more than a dash of fear thrown in—were suggesting an alternative route. Commercial flights within the Democratic Republic of Congo, an area as large as Western Europe, were sketchy at best and definitely not recommended, so a trip that might leave me stranded in Kinshasa, far from the major conflict areas, made little sense. Somalia had assumed the mantle as the world's most dangerous place. Along with international piracy, armed gangs had reduced the capital to rubble; rival clans fought an ongoing turf war whose main casualties were civilians. The only booming business onshore, aside from drugs and guns, was kidnapping. A Somali journalist from Ottawa who had set up a radio station in Mogadishu called HornAfrik had been murdered, and a female journalist from Alberta was being held for ransom. Four British Somalis who had returned to the country to set up an English-language school in Beledweyne—the hometown of Shidane Arone—had been summarily executed by Islamic extremists.

In need of advice, I consulted Jonathan Manthorpe, currently the *Vancouver Sun*'s international affairs columnist. He had served as foreign correspondent for Southam News in Africa for many years and was the first Western journalist to fly into Mogadishu after the fall of the Mohamed Siad Barre regime in 1991. Stroking his well-trimmed, greying beard, he sized me up with the amused intensity of a sage. "First advice? Keep moving. Don't stand around looking lost or confused. An injured animal is fair game in Africa." Jonathan described flying into Mogadishu, not the main airport but a nearby airstrip favoured by drug lords and their couriers, which was considered safer. The cost of protection—you had to hire a technical, a pickup truck with a mounted machine gun and several armed guards—had spiralled when the major news agencies started covering the war. Security at Mogadishu airport had been taken over by a vicious and unscrupulous gang known as the Airport Clan. Jonathan recalled a German colleague who, nose in the air, had turned down the services of these local thugs only to be deliberately shot dead ten minutes later as he left the airport in the back of a rival pickup. By the time our meeting ended, it was clear that I'd have to reconsider not only my starting point but also my final destination. If Somalia was too dangerous, I might at least visit the breakaway Republic of Somaliland, the former British protectorate that, having given up its newly acquired freedom to become part of Somalia in 1960, had reclaimed its independence thirty-one years later in response to the travesties of the Siad Barre regime. The republic was also home to two of the most important poets in the Somali language, Hadraawi and Gaarriye, both of whom I hoped to interview.

I resolved to start my travels in Rwanda and Uganda, the latter a one-time British colony, the former a place where English was increasingly spoken. From my base in Africa's heartland, I could move west into the Democratic Republic of Congo, then north and east into Sudan, Ethiopia and Somaliland. I would take Jonathan Manthorpe's advice and make as many contacts on the ground as possible before I left, to keep from wasting my time and to avoid getting shot.

But first there were other shots to consider. The doctor at the travel clinic in Victoria rubbed his hands and ushered me into his second-floor office across from the shopping plaza, looking as if he had struck gold. Flipping through a chart and checking off each section, he seemed intent on immunizing me against every conceivable ailment except plague. I felt myself growing weaker by the second. The doctor glowed as I recited my itinerary. I had imagined Ebola, sleeping sickness and malaria, but he had a much larger repertoire in mind: new and old tuberculosis, dengue fever, flu, typhoid, yellow fever, meningitis, diarrhea and two kinds of hepatitis. I considered asking if any of his clients had ever run out of the office during the consultation process. Instead, I made a joke about him providing the medical equivalent of a suit of armour.

"Exactly," the doctor said. He unwrapped a surgical kit with two types of thread for closing a cut with stitches: light thread for the face, heavier thread for the body. "And don't feel embarrassed to ask them to use your own needle." None of the ghastly scenarios that came to mind featured a benign "them" within miles. When I got up to leave, he was still scribbling notes on a piece of paper.

My wife, Ann, peered at me with raised eyebrows over her cup of herbal tea as I described the visit. "You mean he didn't offer you an AK-47?"

I finished the preparations for my trip to sub-Saharan Africa, sending off several dozen e-mails to the contacts I'd been given. When it was time to go, I placed a small envelope of instructions on my desk, including a revised will. I'd spoken to each of my daughters the previous night by phone. I picked up the backpack and small carrying case I'd assembled and lingered over a sad and difficult farewell with Ann. I had with me on my computer the image of Shidane Arone's bludgeoned face, photographed by one of his torturers, the perfect symbol of what the West had been doing to Africa for almost two centuries. That photo would help to clarify my purpose and steel my resolve.

two *Orphan and Worm*

I WAS EN route to Kigali, by way of Entebbe International Airport in Uganda. My research had unearthed an irresistible fact: that Entebbe was the final point of departure in Africa for Princess Elizabeth when, after hearing of the death of her father on February 6, 1952, she returned to take up her new life as queen of England and its colonies. The official telegram never reached her at the hunting lodge in Kenya; in fact, she did not hear the news until the BBC broadcast was relayed to her in the nearby Treetops Hotel. Her return home via Sunyani and Entebbe airports was delayed by a thunderstorm. I remember that day so clearly. I was not yet twelve years old, oblivious to the wider world, and had just crossed the intersection at Fourth Avenue and Commercial Drive in Vancouver on my way to Grandview Elementary School when a tall policeman put his hand on my shoulder and announced that the king had died. I must have had a blank look on my face, or been daydreaming, for he went on to explain that this momentous occasion meant school was cancelled for the day. I ran home and exploded into the kitchen, grinning from ear to ear, to announce the news to my stepmother.

"No school, no school! The king is dead!"

My stepmother was not much older than Elizabeth, who was assuming the throne at age twenty-five. I don't know which was more of a shock to her: the king's death or my irreverent behaviour.

If the passing of George VI did not go unnoticed in my family, it was even more significant news for liberation movements in Africa already gathering momentum for independence and an end to empire. India had shed the yoke. Colours and boundaries were shifting on the postwar maps, which included the recently created State of Israel. And it was Israel—for its original displacement of Palestinians, and then the prolonged occupation of their lands after the Six Day War—that would bring the small lakeside settlement of Entebbe once more into the news, in a daring raid that galvanized world attention.

I HAD BOARDED my connecting flight to Uganda at London's Heathrow Airport. As we flew south over France and Spain, my mind was full of images of that famous raid and gun battle at 2300 hours on July 4, 1976, when a team of elite Israeli commandos, faces masked, rolled out of Hercules C-130 transport planes and drove at high speed in a black Mercedes and several Land Rover vehicles across the tarmac to storm the Entebbe airport terminal. The rescue mission had been set in motion on June 27, 1976, after flight 139—an Air France Airbus A300 en route from Tel Aviv to Paris with 258 passengers and 12 crew on board—was hijacked in Athens by two Palestinian and two German militants and diverted first to Libya, then to Entebbe airport. Ugandan president Idi Amin, sympathetic to the Palestinian cause, lent ground support to the hijackers. After the non-Jewish passengers on flight 139 were set free, the hijackers threatened to kill the rest if their demand for the release of fifty-three Palestinian prisoners, forty held in Israeli jails, was refused. The secret Israeli rescue mission flew south all day below the radar along the Red Sea, through the Rift Valley and then west across Lake Victoria. In the bloody gun battle that took place during an audacious night raid lasting less than one hour, all of the hijackers were killed, along with three hostages and one commando, and another ten hostages were wounded. While rescuing the hostages, securing the airport and refuelling, the commandos killed at least thirty Ugandan soldiers and destroyed eleven Ugandan MiG-17 fighter planes on the ground. Idi Amin expressed his displeasure at the invasion by ordering the murder of flight 139 passenger Dora Bloch in her hospital bed in Kampala and the slaughter of hundreds of Kenyan nationals residing in Uganda, whom he regarded as enemy collaborators. Four years later, a bomb exploded on New Year's Eve beneath the dining room of the Norfolk Hotel in Nairobi, killing thirteen people, belated payback against the hotel's Israeli owners for the Entebbe incident.

In a few hours, I, too, would fly in over Africa's Great Lakes region, once thought to be the source of the Nile and the meeting place of Stanley and Livingstone. The contrast could not have been greater between the cordial, highly publicized meeting of the Welsh-American journalist and the famous Scottish explorer, and the violent, clandestine encounter between Israeli commandos and Palestinian hijackers, but the two events had one thing in common: both were a form of high-stakes international political theatre being played out on African soil. The stakes in Africa now were no lower, though the site of violence had shifted from Entebbe to northern Uganda, neighbouring Sudan and the Democratic Republic of Congo, where the ethnic card was played to mask what were essentially land and resource wars, a scramble to control the traffic in oil, diamonds, gold, copper and coltan that was rapidly unravelling the achievements of civil society.

I reclined in the seat and closed my eyes for a moment, trying to control my anxiety. My seatmates on either side were a young African businessman in the window seat wearing an expensive three-piece suit and reading the *Times*, which he had folded in that clever English way of making narrow vertical strips several columns wide, and a middle-aged African woman reading a Bible with the aid of granny glasses perched on the end of her nose. She had a long scar on her right cheek that I took to be a tribal marking. While I struggled with my fears and the claustrophobia of being wedged in the middle seat, the possibility of idle chat or religious testimony posed a greater menace. I thanked the gods of aviation that my companions had serious reading material to distract them.

Once I had quieted my nerves, I had time to go over my notes on the week I'd spent in The Hague two months earlier visiting the International Criminal Court. Violence, human rights abuses and the miscarriage of justice are problems without borders, and no country is immune. The disproportionate number of African Americans in jails in the United States and First Nations individuals incarcerated or living in squalid conditions in Canada indicates that racism is alive and well and that injustice thrives in democratic countries. I'd seen justice miscarried yet again when the Commission of Inquiry into the Deployment of Canadian Forces to Somalia was discontinued at a key juncture, and Private Kyle Brown, a minor and reluctant participant in the killing of Shidane Arone, became a scapegoat for the guilty officers and a dysfunctional military hierarchy. To learn more about international justice and its response to war crimes and crimes against humanity, I'd read Erna Paris's meticulous history *The Sun Climbs Slow: Justice in the Age of Imperial America*, in which she traces the development of Western concepts of justice and the colossal struggle to put in place legal mechanisms to administer justice at the international level. Establishing the International Criminal Court was a gargantuan task, but the principles and procedures were hammered out at the Rome conference in June and July 1998 and ratified by sixty nations on April 11, 2002. The ICC officially came into being on July 1, 2002, with the United States, China, Israel, Iraq, Libya, Qatar and Yemen in opposition.

I needed to know how witnesses are selected and protected by the ICC and how international justice, in a remote corner of Europe, could possibly change things on the ground in Africa. I wanted someone to convince me that George Orwell was wrong when he wrote, "All kinds of petty rats . . . are hunted down while almost without exception the big rats escape." I was also keen to learn how international justice might be perceived by African victims and their families. For example, how would the conviction of Thomas Lubanga Dyilo, a Congolese rebel leader facing charges of war crimes and crimes against humanity, including the forced recruitment of child soldiers, benefit a child in South Kivu who was still suffering from the trauma of being forced to kill his parents? How would such a conviction benefit a raped woman in need of fistula surgery, rejected by husband and family and still struggling to cope with her severely altered state? And what about the shadowy Congolese and Ugandan officials and businessmen who enlisted and supported Lubanga? Would they go unpunished? Already, the case against Lubanga was in danger of being thrown out of court because of procedural irregularities. Chances were good that he would go free, an insult to his presumed victims and a great embarrassment for the ICC, which was facing public criticism and struggling to prove its effectiveness.

The road to the International Criminal Court can be as brief as zooming in from outer space on Google Earth, or it can be an extremely long and drawn-out process, thanks to procedural matters, delays, appeals and the difficulty of accumulating reliable witness reports. Radovan Karadzic, the so-called Butcher of Sarajevo, had managed to evade capture in the Serbian capital of Belgrade for more than a decade by growing a Whitmanesque beard and posing, with all due irony, as a practitioner of holistic medicine.

In spite of its glass and white marble exterior, the ICC is an imposing structure with bars and electrified wires discouraging uninvited guests. *Don't even think of it*, the architecture seems to say. It didn't take long for the two security guards posted in the outer entrance to place me in the unwelcome category, especially after I confessed that I did not have an appointment. I'd forgotten my list of names and phone numbers. When the doctor-professor-visitor-from-abroad routine didn't work, I dredged up a name from memory. "I'm here to see Claudia," I said. "Perdomo?" one guard prompted. I nodded, though I'd forgotten Claudia's last name from the brief e-mails we'd exchanged weeks earlier. I made myself at home in the locked lobby, amidst a few utilitarian chairs and an ATM. Cash, in case someone had to bribe his way into or out of this fortress of international justice? Just as I finished replenishing my modest stock of euros, the door opened and Claudia Perdomo materialized.

Claudia ushered me through the next set of locked doors, and we fetched drinks before selecting a table by the window. The cafeteria at the main entrance served as a reminder that this was a place of serious business: no art, no frills, no lounging about. Claudia told me she was Guatemalan and had worked for the UN on human rights issues. Now she was head of the Outreach Unit of the Public Information and Documentation Section. She explained that her team worked directly with victims and potential witnesses to help them understand the aims and procedures of the court and to create an environment in which they felt safe to share their stories. Her department used various methods to get out the ICC message: posters, booklets, radio programs and something called *un club d'écoute*, a listening circle for communities where there may be only a single radio. In addition to town hall meetings, where videos were shown and response from the victims and their families was encouraged, much use was made of what Claudia called socio-drama, where traumatic events were enacted in front of a group of victims.

"Theatre of the Oppressed," I suggested. "Paolo Freire and all that?"

"Yes, yes," Claudia said. "Freire, politicized language. And art as an enabling process. Sometimes the witnesses make up their own scenes to enact. Sometimes we help them along by providing a very limited script." I knew a bit about drama therapy. It is a powerful medium, but also very volatile for the kinds of emotions it can arouse or release.

I told Claudia about *The Man We Called Juan Carlos*, a self-reflexive documentary made in 2001 by my friend David Springbett with the help of his wife, Heather MacAndrew, that examines the pitfalls of humanitarian intervention. As a young filmmaker, David had gone to Guatemala with Oxfam America to make a film for CBC/*Man Alive* that looked at questions of short-term aid versus long-term development in the aftermath of the devastating 1976 earthquake. An American NGO, World Neighbors, had worked with a small community of highland Maya to help them learn how to improve their yields of corn. In a few years, the community had become almost self-sufficient and relied less on picking coffee for cash on the large ranches or estates. Classes for women in health and nutrition followed. "Each one teach one" was the credo. A young Mayan father named Wenceslao Armira, also known as Juan Carlos, went on to work with neighbouring communities that were also struggling with subsistence agriculture. It was grassroots development in the best sense, but inevitably all these changes pushed up against entrenched power structures, especially the big landowners, who preferred their potential labour force poor and uneducated. Juan Carlos, who had started to use David's 1976 film as a teaching video, was fingered as a troublemaker to be eliminated. As political repression intensified, he joined one of the guerrilla groups and fled to Mexico. His children were killed by army death squads while he was in exile there, and his wife never forgave him.

Unsparingly, *The Man We Called Juan Carlos* examines the culpability of the filmmakers in these subsequent events.

Claudia shook her head slowly. "We need sensitive people working in this area." She glanced at the clock on the wall, then ripped a page from her notebook and wrote down the names of her Outreach colleagues in four of the countries to which I was heading. Her handwriting was bold, the spacing generous. I was glad to find myself included under her protective umbrella.

My host in The Hague, Jane Warren, suggested after dinner one evening that we bicycle over to see Scheveningen prison, where those charged by the ICC and other international tribunals with war crimes and crimes against humanity were incarcerated for the duration of their trials. It was a short hop from her house on Kanaalweg to the unimposing compound with its antiquated castle gate, turrets and gun slots. Two small kids on skateboards, in the company of their father and a brown dog, were cavorting in the parking area. The renovated prison, once used as a lock-up by the Nazis for Dutch resistance fighters, was anything but primitive inside and would have put local hotels to shame, with its 160-square-foot rooms equipped with coffee maker, desk, radio, bookshelves and satellite TV. Detainees also had the use of library, gym, recreation room, flower garden, prison shop and family visitation rooms. Security appeared lax, as the prison door was ajar and no one seemed the least bit interested in the fact that I was taking photographs through the opening. In the UN wing behind this wall, Slobodan Milosevic had been interned and died before he could be convicted of war crimes, defying the court and causing no end of dismay to his surviving accusers. For all I knew, Bosnian-Serb leader Radovan Karadzic, recently apprehended, might have already taken up his lodgings here, with or without his beard. Was he chatting up Thomas Lubanga Dyilo in the courtyard, enrolling in guitar lessons or playing Ping-Pong with Jean-Pierre Bemba, a popular Congolese militia leader and presidential contender charged with crimes against humanity?

A COUPLE OF hours had passed since leaving Heathrow. My religious seatmate had dozed off and was emitting a delicate snore, the open Bible page-down on her lap. The dapper gentleman in the window seat on my right had set aside the *Times* and selected a movie that appeared to involve some sort of vendetta. I was determined not to be lured from my reading by high-altitude misadventures on a diminutive screen, but the changing colours and movement in my peripheral vision kept sucking me in. And the fact that the film was about the search for justice, however self-administered, made it difficult to ignore.

During my time in The Hague, I had met with a French national named Patrick, who worked in the court's public relations sector, and two more women in charge of departments at the ICC. While he gave me a tour of the facilities, Patrick told me how much he had appreciated meeting Simone Weil, the famous writer and Holocaust survivor who had decided to lend her name to the work of the ICC. As Patrick explained, Weil was not impressed with, or convinced by, the notion of international justice, which she considered occasional, random and, at best, unreliable, but she had strong feelings about the rights and protection of victims.

When Fiona McKay found us outside the elevator, she was running late. She directed me to a seat in the visitors' lounge, so I knew our encounter would be brief. McKay had practised law briefly in the U.K., but was now head of the ICC's Victims Participation and Reparation Section. We discussed the differences between courts using common law and the proceedings of the ICC, where witnesses are not cross-examined to evaluate and possibly dismantle their testimony. Like Claudia Perdomo, McKay had a team in Africa who interviewed victims and potential witnesses, using similar techniques of town hall meetings and role-playing. I asked if, given the unreliability of memory, this latter practice might not contribute to something like false memory syndrome, in which a patient too readily accepts a psychiatrist's suggestions about the source of distress.

Fiona knitted her brows and leaned into the question. "That's not my mandate. I want you to understand the nature of witness participation. For one thing, witnesses in this court are voluntary, not mandatory." She explained in detail Regulation 81, which concerns effective participation by victims and witnesses and programs designed to provide them with legal counsel and financial assistance. Given the huge number of victims involved, I suggested, and the extreme nature of their losses, reparations could hardly be more than token. Add the legal, financial and psychiatric implications of bringing witnesses and their representatives all the way to The Hague . . .

Fiona glanced at her watch. "I'm sorry," she said, "but I have to make a pre-arranged call to my team in Uganda in ten minutes." The interview was over, but she invited me to continue the conversation by e-mail. Back in Victoria, I would follow the weekly reports from the ICC that provided updates on various cases and witness programs, but I did not take Fiona McKay up on her offer. However, the contact she gave me with a lawyer named Joseph Manoba in Kampala proved to be the most valuable piece of information I gathered from the ICC.

With only a day left in The Hague, I was anxious to speak with someone from the Office of the Prosecutor at the ICC. My call to Florence Olara, a Ugandan staffer, proved lucky. She arranged a dinner meeting for me with Béatrice Le Fraper du Hellen, a former French diplomat with the intimidating title of director of the Jurisdiction, Complementarity and Cooperation Division, Office of the Prosecutor (OTP), International Criminal Court.

In her late forties, trim, alert and with a disarming candour, Béatrice put me immediately at ease. She was ready to answer all of my questions, even about the two major impediments to the ICC's credibility: time and numbers. Didn't the length of time it took to apprehend suspected war criminals, plus the drawn-out judicial process, with its countless delays and appeals, encourage skepticism about the court's effectiveness? And with so many perpetrators slipping through the net, were the victims' rights to justice ever likely to be satisfied?

"Radovan Karadzic's arrest took thirteen years," I said. "His trial could take several more."

Béatrice beamed. She reached out and touched my arm. "That's the thing, Gary, even after thirteen years, justice will be done. The message is out there, loud and clear. If you commit war crimes or crimes against humanity, you won't sleep well anymore. I love it. I'm very excited about this arrest." She was a true believer, her enthusiasm infectious. "We know the message is getting out. I'm receiving phone calls daily from both the Russians and the Georgians wanting to discuss the situation that is unfolding there. People want to talk to us."

We moved next to the concept of "winner's justice," a term that originated during the Nuremberg Trials in reference to the Allied atrocities that were ignored. How might this affect the credibility of the International Criminal Tribunal for Rwanda, which was convicting former *génocidaires* but ignoring war crimes committed by the Rwandan Patriotic Front, the movement formed by Tutsi refugees to overthrow the repressive Hutu regime? Then there was the question of NATO's possibly illegal attacks in Yugoslavia, where 25,000 bombs were dropped, many of them on civilian targets, killing upwards of a thousand people and causing billions of dollars of destruction to infrastructure, including the Grdelica bridge, which sustained damage when a passing train full of people was hit by two NATO missiles. The International

Criminal Tribunal for the former Yugoslavia (ICTY) had not taken charges of war crimes against NATO leaders and politicians, including Bill Clinton, Jean Chrétien and Tony Blair, seriously enough to open an investigation. Michael Mandel, an expert on international law and Osgoode Hall Law School professor, claimed the NATO attack on Yugoslavia was nothing less than a U.S.-led strike against the authority of the UN.

That led to another question for Béatrice: Does the fact that the U.S. is not a signatory to the ICC make it more difficult for her to conduct her business, especially when so many of the small, troubled states are dependent on the U.S. for military equipment and economic aid?

I was surprised by her candid response. "The refusal of the U.S. to be one of the signatories is, in a certain sense, a problem for us, but it's also an advantage. When we are talking to an unsympathetic African state, trying to get them to help us apprehend a suspected war criminal, they often ask if this action has been prompted by the U.S. And here we can assure them that the U.S. has not only refused to sign but also, thanks to George Bush and his *éminence grise*, John Bolton, set in place The Hague Invasion Act, allowing the Americans to attack this city if one of their nationals is brought before the court. On the other hand, when we are speaking with the Arab states, many of which have close links with the U.S. and its foreign policy, it's useful to be able to say that the Americans are more rather than less onside. So, a certain amount of ambiguity can be put to good use."

By now we'd moved to another table and ordered dinner from a young waiter who kept apologizing for not being able to speak perfect French with Madame. I shared his embarrassment, though my dinner companion seemed perfectly comfortable in English. Justice was being done, at least to the bottle of Pinot Noir.

We talked for a while about what had led Béatrice to the ICC. Championing human rights was a sure way to end your career as a lawyer in France, she told me, especially in government or corporate circles. And burnout? There were days when the strain of dealing with such weighty matters was intense. In Colombia, many of the people she met had been victimized, even the well-to-do.

"I'd be sitting having dinner at someone's house and would ask who the woman was in the painting, only to be told it was my host's wife, who had been killed by paramilitaries. Another would confide about a kidnapped child. And the Chileans—can you imagine, after what they've been through—have not signed on with us. There's something incomplete, a sort of paralysis from not facing the legal implications of what has happened there, not demanding justice." Béatrice turned her face away from me to regain her composure.

I wanted to know what Béatrice thought about the role of foreign mining companies in the ongoing violence. In Darfur, she informed me, she had brought together companies in the conflict area and asked what they wanted most. Was it mere profit or a stable society in which to do business? In public, their response was obvious, she said: yes, we prefer a stable society. I laughed and mentioned Madelaine Drohan's book *Making a Killing: How and Why Corporations Use Armed Force to Do Business*, which argues that most of these companies favour instability, especially if it means getting a better deal from a rebel leader waiting in the wings to assume control. That was certainly the case with Laurent Kabila in the Democratic Republic of Congo. Luis Moreno-Ocampo—the fiery and dynamic new chief prosecutor of the ICC, who had come through the turbulent '80s in Argentina with its torture and disappearances—was especially concerned with the "corporate factor," Béatrice said, the role of resource companies in conflicts worldwide. The OTP's team of lawyers investigating war crimes had been allowed to interview victims, rebel militias, even soldiers in the DRC, but their activities were cut short by the government the moment they tried to interview the white managers and executives of foreign companies in the conflict zone.

All this talk of universal justice left me troubled. While I could see many elements in African and Asian societies—including stoning, the chopping off of hands, female circumcision, forced marriages, child labour and slavery—that cried out for change, there was no ignoring the travesties of justice in the West, where money, plea bargaining, evidence tampering and jury rigging often precluded a fair trial, where innocent people could be put away for life and killers walk free. I had come across a telling anecdote about the communal aspects of justice in Patrick Marnham's book, *Fantastic Invasion*, in which he talks about justice in Africa:

> In the eighteenth century King Damel of the Wolofs captured (after a fierce battle) his neighbour King Abdulkader, who had invaded his country on a Moslem jihad and who had announced his intention, for the glory of God, of slitting King Damel's throat. By tradition the victorious Damel should have placed his foot on Abdulkader's neck and stabbed him with a spear. Instead, Damel asked Abdulkader what *he* would have done had he been the victor. Abdulkader gave the traditional account of behaviour and said that he expected the same treatment, and make it snappy. Damel declined, saying that if he made his spear any redder, it would not build up his town or bring to life the thousands who had fallen in the woods. Instead, he kept King Abdulkader as his slave for three months and then, at the request of the king's subjects, released him. This story was cited all over Senegambia as an example of wisdom and justice. Doubtless King Damel's merciful behaviour was exceptional, but it reveals that the indigenous African sense of justice had no need to be bolstered by the Northern legalism that has supplanted it.

CONVERSATION AT MEAL times on international flights is difficult to avoid. There's something faintly ridiculous about stuffing your mouth in such close quarters with fellow humans, hands raised like a squirrel's to negotiate the limited space, and maintaining a strict silence. I was preparing myself to engage with my seatmates when the woman beside me spoke.

"Would you mind holding my tray while I get up? I'm sorry about the timing, but I have to visit the washroom."

I held the tray of unopened items until the woman disappeared down the aisle, then returned it to her folding table. On impulse, I picked up the Bible she'd left on the seat, pleased to see it was the King James Version, which at least delivers its tales of rape, murder, mayhem and redemption with a poetic flourish. The soft, black pebbled leather cover, marbled endpapers and gold-edged pages were also tasteful. I could not resist checking to see what she had been reading, indicated by the position of a narrow linen bookmark sewn into the binding. The Bible opened invitingly, and the soft onionskin pages spread flat in my palm with none of the stiffness and resistance of ordinary paper.

"I see you're not only an avid reader yourself, but also a curious observer of the world at large. And what exactly is the good lady reading?" inquired my neighbour in the window seat who, up to that point, had been plugged into earphones. Damn, I thought, caught in the act.

I replaced the Bible and undid the screw cap on my tiny bottle of red wine while I considered my reply. "Given that I need a drink to cover my embarrassment, it would not have surprised me to find she was reading from Proverbs 20:1: 'Wine is a mocker, strong drink is raging, whosoever is deceived thereby is not wise.' However, the truth is she was reading from Paul's first letter to the Corinthians 13:13: 'And now abideth faith, hope, charity and the greatest of these is charity.' It would be extremely charitable if you ignored my bad manners."

"Andrew," he said, extending a very large hand. I mumbled my first name and nodded. Then, in his plummy Oxford accent, he added: "I did not take you for a religious person."

I managed a poor imitation of a grin, took another sip of wine. "What are the distinguishing characteristics of a religious person? Don't be deceived by these civilian clothes; I could be a plainclothes priest or a terrorist."

"Is there a difference? At least you're well-read, whatever species. *Toujours la manière.* If we're all to die or be pummelled in the interests of virtue, it's nice to know it will at least be done with style."

"I'm a word addict," I confessed. "I quote from cereal boxes, too. Sometimes, if the words are clever, beautiful or in just the right order, they nest in my ear. My wife considers my punning a pathology. I trust you won't disclose my indiscretion to the lady."

Andrew laughed, broke off a piece of bun and dipped it into the remaining chicken gravy on his tray. "Mum's the word," he said. As the lady in question slid into her seat, the conversation was once again submerged in engine noises, a public announcement about potential turbulence and the drinks wagon coming down the aisle. Andrew replaced his earphones and resumed the movie. Our trays were removed, the lady's meal untouched.

I was too wired to sleep, so I spent the final three hours of the flight to Entebbe reading a play by the Nigerian Wole Soyinka called *Death and the King's Horseman* and looking through my dog-eared copy of Chinua Achebe's *Things Fall Apart*, its title drawn from W.B. Yeats's apocalyptic vision of the world in his poem "The Second Coming." Set amidst Nigeria's Ibo tribe, Achebe's novel offers a unique perspective on Caucasian-African relations, showing how foreign priests are advance troops in the process of pacifying and colonizing the "natives." A slave boy, who is offered to atone for the death of a girl from the tribe, is later required to be killed. Refusing, for fear of appearing weak, to exempt himself from the ritual murder of this child, whom he has come to love as a son, Achebe's central character, Okonkwo, brings shame and bad luck upon himself and goes into self-imposed exile. He returns home after seven years only to find the old ways are under threat by the colonialists. When Okonkwo asks his friend Obierika if the white man understands local customs, he gets this reply: "How can he when he does not even speak our tongue? But he says that our customs are bad; and our own brothers who have taken up his religion also say that our customs are bad. How do you think we can fight when our own brothers have turned against us? The white man is very clever. He came quietly and peaceably with his religion. We were amused at his foolishness and allowed him to stay. Now he has won our brothers, and our clan can no longer act like one. He has put a knife on the things that held us together and we have fallen apart." If justice and hope were in such short supply in the fictional world of African writers, I thought, they were likely to be even scarcer on the ground.

I was already familiar with Soyinka's essays and fiction, where it is not so much the "oppressive boot" of colonialism that is held up to scrutiny as irreconcilable notions of foreign justice and native spirituality. Yoruba tradition, I learned as I immersed myself in *Death and the King's Horseman*, requires that when a chief dies, his personal horseman must commit suicide and follow his leader into the afterlife; otherwise, the chief's soul will wander aimlessly and create chaos for the people. In the play, Elesin, the dead chief's horseman, celebrates life to the fullest as he prepares to meet his obligation. However, the British district officer, Mr. Pilkings, intervenes, insisting the ritual is degrading and primitive. The disruptions of community life resulting from this break with tradition are manifold: Elesin is cursed by his neighbours; his son returns from medical school in Europe and, for the honour of his family, commits suicide in his father's place; the father then kills himself in despair, condemning his soul and bringing disgrace to the community. The blame for Elesin's failure to complete the ritual remains ambiguous, a complicated mixture of vanity, cultural misunderstanding, fleshly indulgence that saps the hero's resolve and blind foreign intervention. Pilkings, still in his skeletal party costume and surrounded by the carnage his intervention has created, asks Elesin's loyal wife, Iyaloja: "Was this what you wanted?" She gives him an earful.

"No, child, it is what you brought to be, you who play with strangers' lives, who even usurp the vestments of our dead, yet believe that the stain of death will not cling to you. The gods demanded only the old expired plantain, but you cut down the sap-laden shoot to feed your pride. There is your board, filled to overflowing. Feast on it."

I admired Soyinka's play for a quality it shared with the plays of Bertolt Brecht: *Death and the King's Horseman* was subversive, challenging audiences to rethink issues, question accepted values. I slipped the play into my backpack and removed a collection of interviews, *Conversations with Wole Soyinka*. In discussion with Jane Wilkinson, Soyinka insisted that change begins one individual at a time and that drama is both a healing process and an agent for the radical transformation of society:

> In the black community here, theater can be used and has been used as a form of purgation, it has been used cathartically; it has been used to make the black man in this society work out his historical experience and literally purge himself at the altar of self-realization. This is one use to which it can be put. The other use, the other revolutionary use, may be far less overt, far less didactic, and less self-conscious. It has to do very simply with . . . opening the audience up to a new existence, a new scale of values, a new self-submission, a communal rapport . . . Finally and most importantly, theater is revolutionary when it awakens the individual in the audience, in the black community in this case, who for so long has tended to express his frustrated creativity in certain self-destructive ways, when it opens up to him the very possibility of participating creatively himself in this larger communal process. In other words, and this has been proven time and time again, new people who never believed that they even possessed the gift of self expression become creative and this in turn activates other energies within the individual. I believe the creative process is the most energizing. And that is why it is so intimately related to the process of revolution within society.

In a second interview that caught my attention, recorded fifteen years later, Soyinka was asked about his poetry collection *Mandela's Earth*, in particular a poem called "Cremation of a Wormy Caryatid," which the interviewer suggested was pessimistic. Soyinka objected to the interviewer's reading: "Whether we like it or not, in terms of effecting change art does have its limitations. And I keep emphasizing that recognition of this is not a negative or pessimistic view of art. For me it is a positive one. Certain kinds of artistic production in my society are left to rot, deliberately. It's part and parcel of the persona of a work of art that it is meant to vanish, to be destroyed in order to be able to reproduce itself. This is the organic nature of art." As an example, he offered wood carvings, created in full knowledge of their perishability. "Yes, there is this work of art, and it is quite possible for little termites to eat into it and destroy it. But those termites cannot . . . destroy the *creative essence* that produced the work of art." While he acknowledges the sorry state of the world, a view he shares with Achebe, Soyinka has tirelessly promoted democratic values and shamed the reactionary political forces in his country on the international stage, often at great risk to himself. He never advocates putting down the pen and taking up arms, but continues to condemn the corruption and violence eroding the fragile unity of Nigeria.

Having also spent much of my creative life convinced that writing is a subversive act and that literature is one of the healing arts, I was thrilled to have this brief interlude at thirty thousand feet with Wole Soyinka. I knew my ignorance of the necessary languages, history and traditions would be a liability on my travels, but I welcomed the moral support of Soyinka's company nonetheless.

As our plane descended over Lake Victoria—its brilliant surface illuminated by the moonlight—my seatmate Andrew offered me his copy of the *Times* and some unexpected advice.

"Don't believe everything African writers tell you. Half of them are living in the past, the other half prostituting themselves to impress Western intellectuals. Once you give up your own language, you've lost both your real audience and your integrity. You'll get more accurate information about Africa talking to people on the ground through a translator—farmers, miners, teachers, journalists, cassava merchants and especially women, who do most of the honest work on this continent."

A blanket of moist tropical air enveloped me as I stepped out onto the tarmac, redolent of exotic flowers and decaying vegetation, a primal funk that reached back to the beginning of time. I was here, at last, for whatever might be in store.

I had confirmed but not paid for my connecting flight on RwandAir and was anxious to find the ticketing agents so I could relax, catch a nap and do some reading in preparation for Kigali. I strode the length of the building half a dozen times to stretch my cramped legs, while an airport employee disappeared with my passport and promised to locate the ticket agents.

Despite its centrality and function as a hub for United Nations flights in Africa, Entebbe airport was quiet this morning, a few dozen passengers like myself waiting for connections, two bored Ugandan soldiers making periodic appearances and a solitary cleaning woman barely awake at the handle of her three-foot-wide dust mop. The airport employee escorted me to the RwandAir ticket counter, where I paid my fare by credit card, then went back upstairs again to the departure lounge for the long wait.

I purchased a copy of "*Exterminate All the Brutes*" by Sven Lindqvist, ordered a coffee and a woeful cheese sandwich at the bar, and found a vacant table next to two backpackers from Illinois who were sucking at bottles of Nile beer and debating the relative merits of trekking to see gorillas in Rwanda or Uganda. As I sipped my tepid coffee and worked towards a negotiated settlement with the stale bread and brittle slab of cheese, the young man addressed me. He was wearing hiking boots, jeans and a bright red T-shirt advertising Roosevelt University and was seated beside a backpack so festooned with gadgets and metal bottles it resembled a tinker's display.

"Hey, man, you here for the animals, too?"

I thought of the interviews awaiting me—sexual abuse, mutilations, unspeakable atrocities—and considered a smart-assed rejoinder, but found myself saying, instead: "Nothing quite so clean or so straightforward as animals." He looked puzzled and took another pull at his beer. I said I was in Africa to learn about human rights abuses and how indigenous systems of justice were addressing these matters.

"I read about some of those things in the *Lonely Planet* guide," his companion interjected, leaning across the table in my direction. Head shaved bald, she was similarly attired, except for a nylon photographer's vest with a plethora of stuffed pockets, and appeared to be studying the operating directions for a plastic water filter, the pieces spread out on the table. "Too bad the violence is impacting the animals," she said, looking up at me with very intense blue eyes. "It's bad enough that we kill each other, but we're replaceable. Not so the animals."

I spent the next seven hours on a badly upholstered bench in the main hall of the terminal building, trying without success to sleep, worried about my safety and inadequate preparation for this journey. Scenes from Irvin Kershner's film *Raid on Entebbe*, with Peter Finch as Yitzhak Rabin and Horst Buchholz as one of the hijackers, alternated in my brain with images of Idi Amin from *The Last King of Scotland*. I'd been in tight spots before—on a Taliban visa in Kabul two weeks before 9/11 or facing water cannons and a hostile military in the streets of Santiago—but had always been lucky and had learned to trust my own instincts and serendipity. If your plans don't work out, something unexpected and more interesting is likely to transpire; that was the essence of the pep talk I gave myself as fragments of conversation, boarding calls, discomfort and a full bladder kept me half awake.

When the first call for passengers travelling to Kigali on RwandAir came over the loudspeaker, I made my way to security. I removed my shoes and took the computer from my pack for closer inspection, Soyinka's words still turning in my mind. He'd found his protective spirit, Ogun, also known as the orphan's shield, and had come to terms with both death and the fact that some works of art are more time-sensitive than others. So be it. The backpackers from Chicago were ahead of me in line, their heads not visible behind their enormous backpacks. They resembled two giant ants. The lady with the Bible, also heading to Kigali, waved to me.

"God bless you," she said. "I'm a genocide survivor going home for the first time since '94. And I'm a bag of nerves." Thus the scar, the uneaten meal. I was so embarrassed by my unfair assumptions about her that I did not have the presence of mind to ask if we could meet to talk in the coming days.

As I stood in line in my stocking feet, I felt vulnerable, insignificant and ashamed of my paltry record as a champion of human rights: too much time at the desk, too little in the arena facing the lions. If it were me rather than my luggage passing in front of the X-ray, I thought, the screen would be blank. When my turn came to surrender bags, shoes, belt and computer to security personnel, I was waved through, as if to confirm my sense of being invisible. Only later did it occur to me that I might have benefited from the last remnant of a whites-first colonial hangover.

three "The Body-Odour of Race"

AT SUNSET, which comes early in the tropics, Kigali was alive with people—the buzz of motorcycles, occasional laughter, white teeth in soft, dark faces. Some were out for a stroll; others were queuing for mini-buses to the suburbs and surrounding villages after a long day's work in the capital. I was grateful for the coming of night, which relieved me of the impulse to look into the eyes of every passing individual to see if they bore the mark of Cain, of having committed unconscionable acts. After thirty-six hours aloft and in airports, I'd spent the morning doing chores, picking up laundry soap, purchasing a cellphone and bottled water. I scoured the busy market for a new battery for my wristwatch, which had stopped the moment I arrived in Rwanda. Avenue du Commerce lived up to its name, offering every sort of enterprise: money-changers, unlit dens selling goods from China, Western-style clothing shops run by Muslims, spaces no larger than broom closets that stocked odd combinations of hardware and cosmetics. And, of course, the ubiquitous street hawkers offering belts, shoes, gum and phone cards.

In the forgiving morning light, the streets were awash with colour and an astonishing array of garments: youths in the cast-off and out-of-fashion T-shirts that arrived from North America in huge, square bundles; women in multicoloured wraps and elaborate headgear with floral and geometric designs; and a skeletal figure in a shiny, ill-fitting black suit that gave him the appearance of an underfed clerk from the office of Scrooge and Marley. A middle-aged Rwandan, inspiring a double take on my part, passed me wearing a T-shirt that featured a life-sized photographic reproduction of the muscular torso of a white bodybuilder—a whimsical reversal of the Oreo cookie syndrome.

Although my peregrinations took me up and down hills with well-maintained streets, ornate gardens overflowing with bougainvillea and flowering hibiscus, my head was filled with scenes of carnage and bloodshed. I paused beside the pool at the Hôtel des Mille Collines—immortalized as "Hotel Rwanda" in the 2004 film of the same name— where the manager, Paul Rusesabagina, famously provided refuge for up to a thousand hunted individuals during the genocide. Fear must have been palpable, death a constant companion, hostile soldiers crowding the courtyard, bands of drunken killers roaming the streets. Phones had been ripped from the walls. From the hotel's high windows, those inside saw figures milling at the roadblocks and heard the screams as machetes did their work. This morning, a group of Japanese businessmen at one of the poolside tables were sharing a joke over cocktails.

As I passed the local administrative offices of the International Criminal Tribunal for Rwanda, I recalled the airport taxi driver's observation about growing poverty and unemployment in Kigali, which he thought likely to increase the crime rate. Don't sweat the small things, I'd wanted to say to him; theft is a petty crime, not to be compared to cold-blooded murder. But murder, too, is a kind of theft, the ultimate kind. I realized the driver had been trying to alert me to the widening gap between rich and poor in the city when, moments after leaving the compound of the Canadian consulate, its manicured garden dripping with flowers, I watched a young man stagger to control a dilapidated wheelbarrow overloaded with sacks of cement. The broken axle had been tied up with rope so the wheel, a wooden rim bound tightly with twisted rags in lieu of rubber, ran at a 45-degree angle. While shiny new Toyotas and BMWs roared past, every muscle in the man's arms and neck strained to bursting as he inched his impossible burden uphill.

Rwanda's capital was clean, anally so. The streets were swept constantly, as if an African Lady Macbeth had hired a fleet of aging crones, bent over their ridiculously short brooms, sleepwalking, to scrub the blood from Rwanda's hands. It was important for the country, post-genocide, to put on a smiling face for the world to attract business, tourism and aid. Kigali had no litter, no dog shit. In fact, there were no dogs. During the genocide, dogs were everywhere, feasting on the rotting corpses, so orders had been given to shoot them all. Kigali was not only clean but also orderly, even in the markets, the crowded bus stations.

A full moon—benign, non-judgmental—cast its pale light down over Kigali and its infamous killing grounds, where bodies had been piled high in the streets, men bludgeoned to death, children smashed into the ground or dismembered, women raped, then tossed into wells or latrines, beer bottles thrust into their vaginas. All this had taken place in a mere one hundred days, time enough for three full moons and for extremists to murder 800,000 Tutsis and moderate Hutus. *Lunacy*, I thought, a derivative of the word *lunar*. No use blaming the moon. Jet-lagged, disoriented, I was closer to understanding the origins of madness. Add fear, drugs, indoctrination, alcohol, a government edict that declares killing your neighbour an act of patriotism, and even a sane individual might morph into a killer. *Might?* No madness evident tonight, only faint strains of Vivaldi from a nearby apartment.

KIGALI'S GENOCIDE MEMORIAL Centre, on the slope across the valley, was quite visible from the restaurant of the Okapi Hotel. I'd been avoiding it for two days, busying myself with errands and following up contacts given to me by friends back home. A short walk uphill brought me to the centre, a white plaster and glass structure attractively laid out amidst palm trees, fountains and gardens, replete with a strange assortment of terracotta figures—part animal, part human, part flowerpot. One of them, a piglike creature with ragged clay arms indented to look scaled or hairy, was talking into a terracotta cellphone.

I lingered in the gardens, trying to focus on growing things, not human remains. But even the landscape artist had insisted on lessons to be learned. Each section of the garden was designed to reflect a major period of Rwanda's history: pre-colonial times, with their relative calm; the turbulent years leading up to and including the genocide; and then this period of hard-earned and relative harmony. Reluctantly, I slipped into the building with its maze of rooms and exhibits. The Genocide Memorial is unusual in placing the massacre of Tutsis in the context of other ethnic and political massacres of the twentieth century, including the Armenian genocide, the Holocaust, Pol Pot's killing fields in Cambodia and the systematic executions in Kosovo. My hands shook as I stood in front of two glass cases full of skulls in neat rows, a wall of portrait photos—children's faces dangling from paper clips—images of dismembered bodies and the chilling blown-up photo of a roadblock at which one of the Interahamwe, the Hutu militia, stands with his deadly machete poised for the camera. Nestled amongst the skulls, I could make out a necklace and a blood-spattered identity card of the sort initiated by the Belgians to distinguish Hutu from Tutsi.

Beside the portrait photos of murdered children hung the French and English versions of a text by Catholic theologian Canon de Lacger—"The Hutu and the Tutsi communities are two nations in a single state. Two nations between whom there is no intercourse and no sympathy, who are ignorant of each other's habits, thoughts and feelings as if they were inhabitants of different zones or planets"—all the more shocking to me because his words reiterated Lord Durham's 1839 report on the situation in Upper and Lower Canada as "two nations warring in the bosom of a single state." I knew de Lacger's description of Rwanda was not accurate, that there had been plenty of intermarriage and contact between the two groups and that they spoke the same language and shared most cultural habits and rituals, but the starkness of the declaration pointed to the ethnic blurring that can exist in multi-cultural societies and is so easily exploited by politicians and demigods.

Even the bright sunlight was merciless as I stepped outside. I was not prepared for what awaited me at the bottom of the steps leading to three massive concrete crypts, where the remains of a quarter of a million of Rwanda's victims had been buried, nor the painful intimacy of the first names appearing on that wall:

Patrice
Euphonie
Chantal
Véronique
Emmanucl
Népo
Gasinabo
Mbimburo
Aimable
Anastase
Sophie
Angélique
Issa
Musabyimana
Habiyarake
Hyacinthe
Karangwa
Abdallah
Pacifique
Charlotte
Mukarubega
Stanislas
Hadidje
Solange
Mohamed

Gratia
Umutangana
Mutega
Hussein
Mugorewindekme
Rwakumbirige
Innocent

The sophisticated exhibit upstairs had somehow contained the violence, rationalized and diminished it. These names were red-hot needles in my flesh, each one representing a particular face, a body pulsing with life, tissues brimming with hope. I walked through the gate of the Genocide Memorial Centre, but could not shake off the legion of ghostly presences.

Valentin, a Congolese refugee who had struck up a conversation with me on the mini-bus to the centre and asked for my cellphone number, joined me for dinner that evening at the Okapi Hotel. He had been a teacher in Goma, he explained in French, but was struggling to survive in Kigali, working part-time in the food industry. He told me he'd sought refuge in Rwanda after his wife's parents were brutally murdered by fleeing Interahamwe, who invaded the Democratic Republic of Congo after the Rwandan Patriotic Front's victory. He ran a finger across his throat for emphasis.

"Has the RPF brought peace?" I asked.

"*La paix?*" Valentin looked uncomfortable and shrugged, his body language answering the question.

"What's the future here for refugees?"

"*Rien*, violence will erupt again," he explained. I caught the phrase *baiser adieu l'Afrique*. Finland was his destination of choice, because he had a friend living there, but kissing Africa goodbye would not be easy. The papers and bureaucratic wrangling required to get this under way would cost 550,000 Rwandan francs, about a thousand dollars, which, with a wife, six children and an irregular income, he had no way of raising.

I lifted my glass of beer to have a drink. The room began to spin. I could see several versions of my companion across the table. Was I having a stroke? Would I die in Africa, just as my trip was getting started? Valentin was saying something I could not understand. I closed my eyes. With my elbows on the table, I rested my head in my hands and very gently rocked it from side to side. I heard the phrase *Centre de Génocide* connected with the word *mauvais* and felt a hand on my wrist. Valentin was trying to console me, convinced my shaking head was a response to what I'd experienced earlier in the day. If I tried standing, I knew I'd fall on my face. Besides, I'd never be able to navigate the precarious wrought-iron steps at the back of the hotel that led down to my room in the annex.

Slowly, Valentin came back into focus. He looked alarmed.

"*Je suis malade*," I whispered, afraid I was going to vomit. After several minutes, the room ceased its gyrations, but I could not make conversation.

I bade a solicitous Valentin an early *adieu*, and with legs weak and trembling I managed to stay upright as I crossed the restaurant and descended the set of rickety stairs, past the array of laundry drying on a mishmash of wires strung up between the hotel and annex. I struggled with the key to the door. Once inside, I leaned for a moment with my forehead against the cool plaster of the wall. Books and various travel items littered the bed. I pushed these futile trappings onto the floor, but did not lie down for fear the dizziness would return. I bathed my neck and forehead with a damp cloth. The face that peered out at me from the mirror was ghostly white. Maybe I was suffering from a combination of jet lag, heat prostration and exhaustion. I'd walked from the Genocide Memorial back to the hotel in scorching heat, mostly uphill, carrying my backpack laden with laptop, books and camera. I'd not slept or eaten properly for three days. And the emotional impact of the memorial, as well as two hours of intense concentration trying to grasp bits and pieces of Valentin's disturbing narrative in French, must have overtaxed my nervous system.

The Genocide Memorial was not my first encounter with a mass grave. Once, on a trip to Chile during the Pinochet dictatorship to do poetry readings and conduct human rights interviews, I went with Canadian writers Patrick Lane, Lorna Crozier and Mary di Michele to the General Cemetery to place flowers on the grave of Pablo Neruda, who had died shortly after the CIA-inspired coup of September 11, 1973. As we wandered with our Chilean hosts, Lake Sagaris and Patricio Lanfranco, through the sobering array of graves—elegant crypts for the elite and walls of small compartments the size of mailboxes that housed ashes of the poor and the humble—we saw a patch of untended graves marked with small white crosses bearing the letters NN, *Ningun Nombre*—No Name. Bodies of the "disappeared" had been quickly interred there rather than left to rot in the desert or dropped from helicopters into the Pacific Ocean.

As we bent to examine these graves, the attendant came running with an armed guard, shouting that we must leave at once and threatening to confiscate our cameras. Only then did I notice the spike of a high-heeled shoe poking up through the ground. I nudged Patrick, nodded in the direction of the shoe and tapped my camera. He began to pepper the guard and attendant with innocent questions, which Pato translated: "Why does an important poet like Neruda have an inconspicuous box for his ashes rather than a well-marked grave to honour his memory?" I managed a single photograph of the high-heeled shoe, and back home the experience found its way into a poem called "General Cemetery."

In a book called *Exorcising Terror*, Chilean playwright Ariel Dorfman offers a moving description of a moment in 1998, just after the arrest of General Augusto Pinochet in London, when the split personality that is Chile was acted out in the streets of the capital: "A motley crew of around one hundred students, dressed like medieval buffoons, their faces painted all sorts of colours, several of them on gigantic stilts, were parading through one of Santiago's main streets inviting the public to a Festival of University Theatre, a sort of Edinburgh Fringe here in our nation's capital. I love how they jumped, they juggled, they played the fools, dancing their joy at being alive, taking over the rather staid

Chilean public space with their carnivalesque celebration of art." This exuberant troop was followed by an assortment of mothers, daughters and wives of the disappeared, the tortured and those executed without trial, women who, in the face of prolonged arrests, beatings and humiliation, had been protesting Pinochet's reign of terror and impunity for more than twenty years. As Dorfman explains,

> They were singing quietly down the street, hands locked, photos of their dead pinned to their dresses, reminding me and the other bystanders who were out shopping or licking an ice-cream cone or about to take in a movie that there was an abyss between the rollicking, multicoloured university students who had just careened through these very streets, drumming and whistling, and the unbearable pain of these women who would not forget, a chasm of memory that needs to be travelled and bridged. Chile is a country where something as normal and wondrous as the young delighting in their own energy and merriment is being challenged by a traumatic past that refuses to be buried. A country where we cannot get on with life until the life that was destroyed here has been acknowledged.

Watching this contradictory spectacle, Dorfman hears a passing shopper mutter, "Communist garbage! Liars! *Mentirosos!* We should have killed the lot of them."

The Rwandan past also refuses to be buried. And the ugly sentiment of the Chilean bystander is one shared by far too many Rwandans, who see old class divisions between Tutsis and Hutus being re-enacted or who, filled with guilt or resentment, wish to see the present social edifice crumble. The genocide is a delicate subject to discuss in schools. Identity cards may have been abolished and the words Tutsi and Hutu outlawed in favour of the neutral term "Rwandan," but discontent continues to brew in many sectors, especially amongst the poor. Many still see economic inequality as being linked to renewed Tutsi domination—a situation that will be a source of struggle long into the future.

The expropriation of low-income neighbourhoods and farming communities in the vicinity of Kigali for the ostentatious houses of foreign investors or Tutsi returnees and *nouveaux riches* will not make the struggle any easier.

Still a little shaky the next morning, I was determined to know how justice could possibly be administered in Rwanda under such complicated conditions. Some of the answers were waiting for me right outside the door of the hotel, where a massive protest was taking place. The Germans had just arrested Rose Kabuye, a government official and retired lieutenant colonel in the Rwandan Patriotic Front (RPF), on a French warrant and delivered her into the hands of France's judicial system. She was accused of conspiring in the 1994 assassination of Juvenal Habyarimana, then president of Rwanda, as he returned from peace negotiations in Arusha, Tanzania. Although the murder of the president, whose plane was shot down on its approach to Kigali International Airport, was claimed by the former regime to have triggered the genocide, it is widely accepted that he was killed by extremist elements in his own party, who resented his indecisiveness and planned to exploit the incident as an excuse to proceed with detailed plans to exterminate the Tutsi minority. The Hutu elite, however, still claim to have reacted spontaneously to a history of Tutsi domination and to continued military aggression after independence by Tutsi exiles. Their response was not to engage the aggressors in battle, which would have been legitimate in terms of international law, but to draw up plans for a final solution, the annihilation of some 800,000 civilian noncombatants. Those plans were already in motion: training camps for the killers known as Interahamwe (defined, ironically, as "those who work together"), vast supplies of arms and machetes ordered and distributed, detailed lists of Tutsis drawn up for each community, and gradual incitement of the populace by way of radio and newspapers, demonizing the Tutsis as "enemies" and "cockroaches."

"You have to work harder, the graves are not yet full."

Foreign powers contributed to the genocide with moral, financial and military support. France, determined to maintain its influence in

Africa, supplied arms and training and helped spirit the perpetrators of the genocide safely over the border into the Democratic Republic of Congo, where they have continued their carnage. When, after fifteen years, Rwanda published the names of high-ranking French collaborators and demanded an apology and compensation, a lot of heat was generated in France. Justice was not the main issue in the arrest of Rose Kabuye; rather, it was a misguided attempt to shift attention away from France's complicity in the genocide, a legal proceeding that seemed likely to backfire.

I took a few photographs from the balcony of the Okapi, then went into the street to join the protest march, where I was offered a small circle of cloth imprinted with a white rose on a green background to pin onto my shirt. As far as I could see, I was the only *mzungu* in the parade of thousands that snaked through the city and down the hill towards the German embassy. Most businesses and government offices had been closed for the day, I learned, so there was a festive air to the demonstration, children pushing hoops, women dressed in their finest, men in suits or casual dress, lovers entwined. They all navigated the flower gardens of the Place de l'Unité National roundabout before winding past the Hôtel des Mille Collines and the notorious Sainte Famille church, where in 1994 Father Wenceslas, dressed in jeans and T-shirt and wearing a revolver, had delivered some of his frightened parishioners to their deaths. The mid-morning heat was intense. Bright umbrellas kept sunlight out of the eyes of infants strapped to their mother's backs. To my left, two young women with finely braided hair hoisted a huge sign that read NAZISM IS NOT DEAD. Another sign said: ARRESTED IN BREACH OF THE GENEVA CONVENTION. Noticeably absent were any messages in French, a reminder that the government had recently announced that French would be replaced by English as the second language in the country. Rwandan flags, with their ascending horizontal bars of green, yellow and pale sky-blue with a golden sun, waved gaily; overhead buzzed the international media, filming the demonstration from the safety of helicopters. Outside the German embassy, the crowd pressed close together to see the posters and listen to a man belting out his message from the back of a small pickup truck mounted with amplifiers.

The speaker addressed the crowd in Kinyarwanda, his voice rising higher and higher in an irritating rant, which the young man next to me, wearing a T-shirt bearing the image of Rose Kabuye, explained was a well-known Christian prayer. The speaker's tone made me think of A.M. Klein's poem "Political Meeting," in which Camillien Houde, the mayor of Montreal, addresses a crowd of Québécois during World War II, stirring up hatred for the English and the idea of conscription. A phrase from that poem came back to me: "the body-odour of race."

"What's he saying?" I asked.

"He's encouraging God to save Rose Kabuye," the young man explained.

This made me angry. "Why God would bother with a political matter of this sort," I said, raising my voice over the din, "when he did not deign to save 800,000 people from horrific deaths during the genocide?" Several people turned to look at us.

"The winds of Satan were too strong," my informant said, adding that he considered the RPF God's avengers.

I wanted to shake this stupid fellow, with his trite answers and superstitions. Instead, I let the matter drop and headed uptown along one of the back streets, pausing to catch my breath in the shade of a small tree. It was there I met Sam Nkurunziza, a charming young medical student from Uganda who was freelancing as a journalist for the Rwandan *New Times* during the holidays. Notebook in hand, he confessed to following me up the hill and asked what I thought of the demonstration. I said the rhetoric of such events always bothered me, that the hysteria created by speakers reminded me too much of the overbearing manner of evangelists and dictators.

"In what way?" asked Sam, pen poised to write.

"Their aim is to manipulate, not to inform. Also, they remind me of the fact that the media was used to stir up hatred for the genocide. The new Rwanda would be better served by reasoned argument and cultivated debate than by empty accusations and emotional diatribes."

When I had concluded my own mini-diatribe, Sam assumed a bemused expression and invited me to join him for a coffee at the Bourbon Café and Internet Bar.

Sam was born in Uganda to Tutsi parents who had fled Rwanda, his father a Christian minister. His favourite author, he informed me over lattes, was Sidney Sheldon, especially Sheldon's autobiography, *The Other Side of Me*, a detailed account of his struggles to become a writer while working as a busboy, an usher and store clerk, then training to be a pilot in the U.S. Army Air Corps. I asked Sam what it was about Sheldon's book that impressed him.

"Everything! I want to write my life's story like that," he replied, sucking the foam off his latte, a small puff of which stuck to his nose.

We parted with vague plans to meet again. The next morning, I received an e-mail from a friend in Toronto, letting me know I'd made the news in Kigali. In the online edition of the *New Times*, I shared the limelight with Rose Kabuye and with Tony Blair, who was in Rwanda to open a juice factory, working off post-genocide guilt in his new role as President Paul Kagame's unpaid business consultant.

Not long after arriving in Kigali, I'd gone to the offices of the Gacaca courts in Kacyiru to talk to Denis Bikesha, director of the Training, Mobilisation and Sensitisation unit. Gacaca (pronounced ga-cha-cha), which means, roughly, "justice in the grass," is a traditional Rwandan process for resolving disputes and transgressions. After the genocide, Rwanda's public services, including the judiciary, were in tatters, and the jails were bursting with upwards of 120,000 prisoners, many unaware of the charges against them. In the decade that followed, only six thousand cases were handled by the Rwandan criminal courts; it was obvious an alternative was needed, or most of those incarcerated, the innocent along with the guilty, would serve a life sentence or die in jail without having their day in court.

Denis invited me to accompany him to an official function in nearby Gasabo, where a new justice centre was being opened. We drove in his 4x4 Toyota past small agricultural plots and clusters of houses to arrive at an imposing white structure, totally out of keeping with the almost impassable dirt roads and ramshackle adobe dwellings nearby. The parking lot was full of cars, dignitaries and a smattering of soldiers. I followed Denis as he threaded his way through the audience to a couple of empty chairs near the front of the new hall, where future district sessions of Gacaca would be held. In late 1994, the UN Security Council, acting under Chapter VII of the United Nations Charter, had established the International Criminal Tribunal for Rwanda (ICTR) in Arusha, Tanzania, to try high-ranking officials involved in the genocide. Justice Hassan Bubacar Jallow of the ICTR had flown in from Arusha for today's event and was wrapping up his speech at the podium. He talked of putting an end to impunity and praised the government of Rwanda for its efforts to rebuild the justice system. During the reception that followed, Denis introduced me to various government officials and members of Rwanda's Supreme Court, immaculately turned-out men and women speaking English, French and Kinyarwanda. On the spot, I decided not to follow up any of these connections. I wanted to see justice at work for myself.

On the drive back, I asked Denis if Gacaca resembled the Truth and Reconciliation Commission in South Africa, where confession is sometimes followed by forgiveness. "No," he said, "we expect punishment to be meted out for crimes, appropriate punishment, which takes into account time already served." Although I was aware Gacaca had been set up to try the Hutu *génocidaires*, and that no crimes committed by the invading Tutsi army would be heard in these courts, I was surprised to learn that many of the Gacaca judges, chosen from amongst the village elders across the country, were illiterate or poorly educated and had to be coached in the basics of justice. Denis admitted that 25 percent of them—40,000 of the 167,000 lower court judges appointed—had had to be relieved of their positions because they too were suspected of involvement in the genocide.

When I received permission to attend a Gacaca session in Butare, Sam jumped at the opportunity to act as my interpreter. He could smell a good story. His English was excellent, and his youthful idealism and quirky sense of humour would prove a tonic. But first, he arranged for me to meet his friend Angélique, who he told me was an innocent victim of Rwanda's justice system. The three of us met for drinks in the gardens of Kigali's new Hotel Banana. I spent an hour waiting for them to show up. When they did, there was obviously a bit of tension between Sam and Ange, who was ticked off at him for not phoning to say he'd be late. After introductions, we were shown to a table in a quiet corner of the restaurant gardens, where young bamboo and tropical flowers bordered an attractive jigsaw of slate paving stones.

While we ate dinner, Sam gave a spirited account of his life as a newshound—bouncing back and forth between an ecumenical religious conference and an agricultural interview, trolling the streets for human interest material and racing to the office to file his stories—all the while speaking to mc in English but taking the time to translate into Kinyarwanda for Ange.

Ange was an attractive woman in her early thirties, at present unemployed but interested in starting a small business in Kigali. Unfortunately, Sam had not explained to her in advance that he wanted her to share her personal story with me. While the dishes were being cleared away and I rummaged in my daypack for my notebook and a small recording device, she gave him such a fierce look I expected her to get up and leave. Sam placed a comforting hand on her wrist, but she shook it off, clutching the wrought-iron rim of the glass tabletop. After an awkward silence, she leaned back in her chair with hands crossed in her lap and agreed to let Sam translate what she said as long as I did not use the tape recorder.

"My father was Hutu, my mother Tutsi," Ange began. She had been smiling and at ease moments earlier, but now her voice dropped to a whisper. Two months before the genocide, while she was still a teenager, the Hutu bodyguard of someone high up in the government had kidnapped Ange and several other girls. He and his friends raped them repeatedly; and one of them stabbed her in the ribs. Her voice, which had become almost inaudible, stopped altogether. Sam cleared his throat, offered a few encouraging words.

With great effort, Ange dredged up more of her terrible story. When her tormentor was killed during the fighting, she and the other girls escaped, fleeing over the border to Goma. When the bloodbath ended and the RPF had taken power, she returned to Kigali, thinking she would be safe. Instead, a man accused her of murdering his two children and, though she knew both children had died of dysentery in DR Congo, Ange was thrown in prison.

"Twelve years," she said, her voice trailing off again. Sam asked her to repeat what she'd said. "Almost a third of my life wasted—and for what? Nothing." Just when I thought the story was over, that we had taxed Ange beyond her capacity to speak, to recall, she explained that twelve years into her prison sentence her accuser finally confessed that he had blamed her on the basis of hearsay—the testimony of someone who coveted her family's property. Ange was released. Her sole consolation for all those lost years was that she had fallen in love with another prisoner, also falsely accused, whom she would later marry.

Darkness had settled around us. Uncertain shadows, set in motion by flickering candlelight, performed an agitated dance. I shifted in my chair. Although the temperature had dropped, I was sweating profusely and had difficulty meeting Ange's gaze when, finally, she looked up at me. I asked how she had coped with her anger at the perpetrators and her resentment against the government.

"I had to put it aside," she said. "It was poisoning my life, making me ill." She looked off into the shrubbery. "It would have killed me by now." As she spoke, the fingers of her left hand moved to touch the spot on her ribs where she had been stabbed, fourteen years earlier.

As Sam and I set out two days later on a crammed mini-bus for Butare at 6 AM, I had Ange's story very much in mind, particularly the question she'd asked him following our meeting: "What does Gary want?" The thin veil of fog was lifting from the undulating hills and terraced slopes, its damp residue heightening the chorus of greens that greeted the early riser. Already, workers crowded the roads, some emerging from the fields, only their legs showing beneath loads of thin green stalks that came down to their waists like the exaggerated hair of demons in an Asian drama. I'd brought along fruit, cookies and water, which we consumed during the two-hour trip through small towns and villages. As we passed one genocide memorial after another, I felt uneasy for having intruded on Ange's privacy, upsetting her delicate balance. I was embarrassed, too, to admit I did not know exactly what I wanted.

Considered a safe haven during the genocide, Butare instead became a slaughterhouse where an estimated 220,000 men, women and children were hacked to death. Though its name had been officially changed to Huye, the memory of spilled blood lingered. I grilled Sam about ethnic divisions, whether blame for the genocide lay partly with the Belgians for favouring the taller and lighter-skinned Tutsis and for issuing identity cards.

"It wasn't race," Sam insisted, as he dismantled one of the oranges. "It was class. If you had ten cows or more, you were called a Tutsi. If your cows were stolen during the night, you'd wake up in the morning as a Hutu." This sounded pat, ignoring a long history of privilege, especially coming from a man with Sam's prominent nose and lean, handsome face, but I let it pass.

No one we spoke to on our arrival in Huye knew where the Gacaca session was being held. After a trip to the police station, where Sam located one of the judges, we flagged down a couple of moto-taxis and made our way to a stadium at the other end of town. Three women were seated on a viewing platform in the bleachers, where a large folding table and metal chairs had been set up for the judges. Beyond the far wall, I could see an enormous eucalyptus tree and billboards advertising a Bon Marché and something called ECAN Buildings and Public Works. Several men were using handfuls of rock dust to mark wobbly chalk lines for a football game scheduled later in the day. However innocent the task, I couldn't help recalling the infamous soccer stadium in Santiago, where so many were tortured and killed under Pinochet, and the stadium I'd visited in Kabul where public executions by the Taliban regime had become a spectator sport.

At 10 AM, two hours after the trial was scheduled to begin, no one else had arrived at the stadium. Sam and I were about to give up when an old man on a bicycle appeared to inform us and the three women that Gacaca had been shifted to a theatre several hundred yards away. In the parking area around the theatre, a diverse crowd had gathered—VIPs and security personnel, a cluster of prisoners in salmon-pink shirts and knee-length shorts, and a mixed bag of relatives, locals and the simply curious—waiting for the announcement to proceed inside. Just as we found seats, the prisoners were ushered in and seated in the front two rows, followed by the judges, two women and four men, each with a sash of office across the chest. When they had taken their places onstage, the presiding judge requested our attention as he described the function of the court and explained the procedures. This was followed by a minute of silence, meant to commemorate victims of the genocide, but thanks to the open eaves the meditative moment was overwhelmed by shouts, bird calls, portable radios blaring, children crying and motorcycles revving past. The judge interrupted his reading of the rules to answer his cellphone, then invited a man in the front row—dressed to the nines in an expensive tailored suit and seated beside an immaculately turned-out woman in traditional garb—to take the chair designated for the accused.

I nudged Sam. "Who's that?" He put a finger to his lips to silence me and leaned forward. There were no microphones, and the theatre's acoustics were hopeless.

"It's Marcel Gatsinzi, the minister of defence," Sam whispered, eyes wide, face animated. He looked as if he'd won the lotto.

The judge informed the assembly that all of the accused who appeared in court that day were invited to make a public confession and request forgiveness. If this happened, the court was authorized to reduce the person's sentence by half. He took several minutes to flip through his large book to locate this case and read the charges. The minister of defence, as a former army commander in the Habyarimana government, was accused of complicity in the genocide. Somehow, he had slipped through the cracks and risen to a position of power in the new Kagame regime.

As there are no lawyers in Gacaca to speak on behalf of the accused, Gatsinzi spoke in his own defence, denying the charges. Sam translated for me a snide remark directed at the accused by one of the judges: "Were there no butchers in Butare, then?" Through the open door, I could see the rhythmical swing of a gardener's sharp blade through the tall grass.

The minister of defence was a controversial figure. Gatsinzi had been appointed immediate successor to army commander-in-chief Nsabimana, who died in the targeted plane with the president. However, he was not popular with the plotters, who quickly arranged to have him replaced. Gatsinzi was given the position of command in Butare, where he managed briefly to keep a lid on things. But as pressure built to eliminate local Tutsis and the thousands of others who had taken refuge in the city, he was quickly removed from his post and replaced by hardliner Colonel Muvunyi, who set about the business at hand.

The exchange went on for over an hour, the minister proclaiming his innocence, providing alibis, and the judge making what I took to be more sarcastic asides. To my left, the pink-clad prisoners were shifting in their seats, whispering to one another; one of them was taking notes, but when the minister finished and leaned back in his chair, not a sound could be heard in the theatre.

The minister of defence, Sam later explained, had insisted he knew nothing. Because he was married to a Tutsi, Gatsinzi claimed he was kept ignorant of the extermination plans, shifted from job to job and finally to Arusha, where power-sharing discussions with the RPF were under way. The presiding judge called on two witnesses, prisoners who had been brought to testify against Gatsinzi. One had accused him of providing arms for the genocide, but on this occasion both witnesses refused to testify on the grounds that it would jeopardize their forthcoming trials. The court could not decide on Gatsinzi's case, and it was deferred to a later date.

The next accused summoned by the judge denied all charges against him. The fourth prisoner, the one who had been furiously making notes during the proceedings, also denied any involvement in the genocide. The fifth and last accused to be summoned, director of the Kigali prison during the genocide, charged with arming and releasing prisoners to go out and murder Tutsis, insisted he could not be held responsible for what had happened in the prison when he went home at night. Two fellow prisoners stood up to accuse him. Testimony from the accused murderer's son—to the effect that his father hated Tutsis and had killed many of them—was read out in court. To no one's surprise, the ex-warden dismissed the testimony.

"My son must have gone crazy from the trauma."

This comment prompted laughter from the judges and some members of the audience. Offhand remarks from the judges, prisoners playing to the audience, the failure to extract a single confession or acknowledgement of complicity: this was theatre, all right, bad theatre, bordering on farce. I was disgusted.

Sam was ecstatic on our return to Kigali, convinced he'd get a front-page story with the headline MINISTER OF DEFENCE DENIES ROLE IN GENOCIDE.

"Don't forget you're working for a government newspaper," I said. "They'll never run that story." Nothing of the huge debt to victims and their families had been settled today, I pointed out angrily. Gatsinzi was a free man; he could go home to his lovely wife, his expensive belongings and his position of power.

Sam could not be dissuaded. He believed the Gacaca courts were working, that even someone high up in the government could be brought to justice by the people's court. I was not convinced. There might be official reasons for giving Gatsinzi's trial a low profile, keeping it off the radar, I argued.

Over the next three days, Sam would be questioned twice by the editors of the *New Times*. How did you manage to get to Huye for this trial? Are you sure these details are correct? We have to check more facts in this case, the editors told him. A week later, he was still hopeful that his account of the Gacaca session would run.

A VISIT TO the Village of Hope outside Kigali had been recommended by my friends Peggy Frank and Peter Bardon, founders of Positively Africa, a small NGO working on behalf of AIDS sufferers. The route to Kagugu wound through clusters of tiny adobe houses without heat or running water and past the palatial dwellings of the rich, some with swimming pools, fleets of vehicles and extensive gardens. The cab driver agreed to wait for me when we finally located the Village of Hope, several one-storey buildings in an overgrown field. Peninah, the director, waved a welcoming hand from the open gate. After some brief words of greeting, she introduced me to four women seated on grass mats in a small room. They were making the intricate, tightly woven baskets that sold in the organization's showroom for ten thousand Rwandan francs, about twenty dollars. Each woman received the entire amount from a sale, Peninah informed me. Half the cash went to her directly for food, expenses, school fees; the other half went into her account to buy materials.

The four weavers did not know what to make of me, a gangly foreigner appearing out of the blue with notebook and camera. Having no idea how to begin, I spent a few minutes talking about my childhood: the death of my mother when I was seven; my father, a tradesman, a good carpenter and truck driver, but also an alcoholic. I'd grown up poor, I explained while Peninah translated, working at odd jobs, loading boxcars with sacks of sugar, delivering newspapers, stocking grocery store shelves. I felt the falseness of what I was trying to convey even as I uttered the words.

The four women continued with their weaving, slipping more straws into the continuous grass cord being bound with thin black or red strips of what appeared to be plastic. I cleared my throat, uncertain how to proceed. I tried another tack. In Chile, I told them, I'd interviewed women whose family members had been tortured or murdered by the military junta. What amazed me about those Chilean women, I said, was their determination to tell their stories, however painful, in the hope that justice would be done and the perpetrators made to pay for their crimes. They had repeated their stories for me, a complete stranger, at great risk, in the hope that I would pass them on to the outside world.

"So, if one of you would like to share your story with me," I concluded, looking as if I'd been sucking a lemon, "I'd be very grateful."

Just when it appeared my efforts had failed, one woman began to speak, her face expressionless, her eyes averted.

"My name is Louise," she said, waiting for Peninah to translate. "I was touched by the story of your mother dying when you were young. This happened to me, too. My mother was killed trying to catch a bus when I was just an infant; I never knew her at all. I don't even know what she looked like. My father left the family to find work and never came back. My uncle tried to look after me, but he resented the burden. I was sent to the church care centre. When I was a little older, they tried to send me back to my uncle, but he did not want me. The church gave me some work with the sisters. I was put into a group of young people, but the anger was too much. It felt like a stone in my heart, so I left."

Louise paused. She was wearing a pale grey blouse with tiny multicoloured dots, a skirt with orange and black geometrical lines and matching headscarf. She could have been in her late thirties, but looked much younger. As she resumed her story, I listened to the rise and fall of her voice and watched emotions shifting like a weather system across her face.

"When other kids talked about their families, I felt angry, because I never knew my parents." She waved a handful of straw back and forth, a frail banner of protest at the injustices in her life. Anger had driven her out of three different groups, she said. Eventually, she was allowed to teach primary children, which proved to be a healing experience. During this time, she fell in love with another teacher who was also an orphan. They married secretly and moved away, telling no one. When a child did not come right away, she was convinced God was punishing her. After she gave birth to a daughter, then a son, things seemed to improve; her anger began to ebb. But it did not last.

"My husband was killed during the genocide. I was raped and became HIV positive. I still blame myself for disobeying God."

As Peninah translated, I made some hasty notes, trying to hold onto the details of Louise's story. The other women carried on with their weaving, occasionally nodding, glancing at me briefly. I felt like an intruder in that tiny cubicle with its bare walls and wished I'd chosen to sit on the floor rather than a chair.

After the rape and the murder of her husband, the stone of her anger grew heavier, Louise explained. She thought it would destroy her. She married again, to a man whose wife had been killed in the genocide, and had two more children, but then that husband also died. Just when it had seemed the stone would finally crush her, someone told her about the Village of Hope. Much of what Louise said to me was lost in the translation, but I could see her relaxing and becoming more animated as she neared the end of her story, as if the weight of memory itself were lifting.

"My children sometimes go to bed hungry," she said, by now looking me full in the face and smiling. "But I have this place, the love of Peninah. And I have friends to help with my sickness."

As I stood to leave, from somewhere in the compound the sound of singing rose, flooding our tiny pocket of intimacy, high, clear notes washing away grief and pain, sending up a message of hope. I could still hear it minutes later as my taxi reached the top of the hill.

Before my flight to Uganda, I'd make a second visit to the Village of Hope, where more painful testimonies awaited me and where each weavers had selected her best basket for me to buy, a joyous encounter that would prompt more singing and dancing. Word of my visit and the impending purchase had spread, so a fifth mother had joined the group.

I had expected to have more answers from my visits, but was burdened instead with a host of new questions relating to violence, justice and human rights. What would constitute justice for Louise and her friends? How were they expected to relate to the slow pace of justice in a remote suburb of The Hague?

The International Criminal Tribunal for Rwanda in Arusha had been shockingly slow getting under way, impeded by incompetence, inefficiency and underfunding. Most Rwandans remained oblivious to, or poorly informed about, what was happening there. However, when the court decided in 1999 to drop charges against Jean-Bosco Barayagwiza, considered a major player in the genocide, on the grounds that the "rights of the accused" had been violated, all hell broke loose. The decision so outraged the Rwandan government that it severed connections with the Arusha tribunal and refused to cooperate in the gathering of evidence. Courageous women had saved the day. Even though they knew their personal grievances might never be addressed, that they would have to content themselves with symbolic, rather than specific, personal justice, one by one they placed their safety in jeopardy and testified against the accused to the point that Swiss chief prosecutor Carla del Ponte, a dynamo who had previously been involved in a legal battle with the Russian mafia, was able to convince the court to reverse the decision. This restored the court's credibility in the eyes of Rwandans, but did not help its reputation within the legal community.

As Elizabeth Neuffer explains so movingly in *The Key to My Neighbor's House: Seeking Justice in Bosnia and Rwanda*, the "arithmetic of justice" in Rwanda defies any simple solution. After standing ankle-deep amongst the bones and scattered garments of Tutsis massacred in the Ntarama church, Neuffer had to acknowledge that "the sheer scale of Rwanda's atrocities confounded conventional efforts to bring justice. It was one thing to pick up the bones of the dead and arrange them in rows for the viewers at the genocide sites. It was another thing to bring all killers to justice."

ASIDE FROM THE polite, persistent youths trying to sell maps and handmade cards, I had felt neither harassed nor threatened during my time in Kigali. So it was a shock one afternoon when, crossing a busy intersection between buses, I collided with someone and felt fingers reaching into my shirt pocket. Instinctively, I knocked the hand upwards, swung one foot behind the pickpocket's knees and pushed. He went down like a rock. He lay there for a moment, listening to my string of accusations, then fled, his lean, earnest face disappearing into the crowd. I felt embarrassed by my violence and my absurd, bullying outburst, all for the sake of some poorly stowed petty cash. As I gathered myself together, straightened my clothes and rounded the corner, a bus driver who had observed the whole debacle leaned out of his window and shouted: "*Mzungu*, bravo!" He gave me a thumbs-up, then added, "You are a man!"

I preferred a different, less aggressive version of manhood, and I was about to encounter a very good example. Jean de Dieu Basabose, who runs an organization called Shalom, Educating for Peace, had arranged to meet me at the nearest mini-bus terminal, a warren of private carriers competing for business. From there, we would set out for Rwamagana, where he was scheduled to give a pep talk on peaceful sport to a group of young soccer players. Jean de Dieu had completed a degree in peace and security studies in South Africa and was dedicating his life to the promotion of non-violence. With a Hutu father and a Tutsi mother, he considered himself a product of, and perfect symbol for, the love that is possible between ethnic groups. Earlier, he had invited me to listen to a radio broadcast he was giving on a religious station in Kigali. Thinking he would deliver a pat homily about Christ the Prince of Peace, I was chastened to find him quoting from two provocative essays. From the copy of the talk he'd made for me, I learned that one essay, by Brian Martin, was "Non-Violence versus Capitalism," an intelligent Marxist critique of material interests and the violence required to keep them alive; the other, by Jean de Dieu's former professor Geoff Harris, was called "The Case for Demilitarisation in Sub-Saharan Africa."

As we sped east along the highway, our mini-bus dodging pedestrians and passing slow trucks on perilous curves, Jean de Dieu pointed out the ritual palm fronds that had been laid out for a visit by the Kenyan head of state the previous day. He had nothing good to say about the Rwandan government's takeover of many poor residential and agricultural parcels of land for industrial development and high-end housing projects.

"Expropriation is the national theme song these days," he said.

"What about consultation?" I asked. "Is there no way to resist or appeal this process?"

"The word *consultation* does not appear in the Rwandan dictionary."

In Rwamagana, a sprawling village with dirt roads, Jean de Dieu introduced me to the coach and the players, aged nine to twelve, drawn from local schools. I watched the boys play a game, impressed by their skill, then joined them to share the drinks and a large carton of cookies we'd brought along. The playing field bordered a camp for detainees, men sentenced to community service as a final payment of their debt to society. It was a felicitous juxtaposition, I thought, young boys being trained to play fair alongside the thin wooden lock-up where some of their fathers, brothers and uncles were doing time for crimes committed during the genocide.

On our way back to Kigali, we stopped at a small compound for rescued street kids. Les Enfants de Dieu was a unique project funded by an organization in the U.S. and run by a young Rwandan named Rafiki. Les Enfants de Dieu had purchased enough land to operate a small fish hatchery and to raise ducks, rabbits and vegetables, but the most interesting feature of the project was its structure.

Rafiki took delight in explaining how the system worked, how it helped build confidence and reintegrate the youngsters into society. The kids elect seven ministers from among their ranks each year to lead unions and departments. Every important decision is made by the kids themselves, all of whom serve on various committees, including finance, labour, entertainment and social affairs. When his computer broke down, Rafiki told us, he had submitted a request for repair or replacement to the "minister" of finance. The young lad who held this position had invited Rafiki in to discuss the matter, then turned down the request on the grounds that the expense would jeopardize the community's food supply for the month. Another of the boys, a precocious fifteen-year-old, had convinced a local radio station to let him host a program of interviews with business leaders, street people and politicians. During one program, the boy asked a young drug addict about his vision of the future. When the addict explained that he wanted to be rich and own a car, the interviewer interjected: "It seems you are planning to die young." Weeks later, after Rwanda's actual minister of health had laid out her program of action for the listening audience, her young host and interrogator said, with due deference: "Your plans are good, but they are identical to the unfulfilled promises of your predecessor. What's your secret?"

Back at the hotel, where a cool breeze accompanied the dying of the light, I spent an hour on the computer, checking e-mail and making notes. Jean de Dieu's own secret was a deep religious faith, something I had never quite achieved. I envied him his calm but righteous indignation at the injustices in Rwanda and the social gospel and good works that kept him busy.

"BONJOUR, MON DOCTEUR." Valentin, seated across the table, was determined that we maintain contact. He'd been a regular visitor at the Okapi since our chance meeting on the bus to the Genocide Memorial and my subsequent dizzy spell. I had asked to be called by my first name, but he insisted on using the academic handle—ironically, I assumed. His visits were a trial, as they required me to speak French, excellent preparation for the time I'd spend in DR Congo, but exhausting. Employed only part-time, Valentin was always short of money and had been obliged to send two of his sons back to Goma, in DRC, where they could attend local high schools more cheaply. As a member of the Neo-Apostolic Church of Rwanda, Valentin at least had the consolation of religion and was quite shocked when I confessed to being a nonbeliever. After several attempts to convince me of the existence of God, he settled on the argument from design.

"When you read a book," he said, tipping back the large bottle of Primus beer I'd bought him, "don't you want to know who the author is?"

"Only if it's a good book."

Valentin was not fazed. "So, when you look at the world around you, don't you want to know who the creator is?"

I had no desire to engage in another long debate about God, the origin of species or the myths these conundrums inspire. Instead, I smiled and granted him his point.

We had been talking about my desire to visit Goma and the recent rebel attacks that had driven another 200,000 people from their villages to camps for internally displaced persons (IDPs) and into the outskirts of Valentin's former hometown. Congolese Tutsi general Laurent Nkunda and his followers, who had captured an army base with its cache of guns and anti-aircraft weapons and created havoc in the region, were the subject of much international debate. Murder, rape, looting and burning by militias had become rampant in North and South Kivu. Nkunda claimed to be protecting the Tutsi population in DRC from attacks by the Hutu *génocidaires*, yet he was accused by the UN and other organizations of receiving financial help, military supplies and personnel from Rwanda and of being anything but a guardian angel. Kagame, Rwanda's president, denied official support for the rebels. Late-night news was full of footage of thousands of displaced people on the roads just north of Goma, moving in both directions. Retreating DRC government forces proved nearly as brutal as the rebels. Uneducated, poorly trained and unpaid, they were known to loot and rape as they fled.

While our views on the subject of God differed, Valentin and I agreed on at least one thing: the fighting in the eastern provinces of DRC was about resources—gold, copper, diamonds, coltan—not tribal conflict, though the latter had been exaggerated and exploited by the government and foreign-owned mining concerns. Neither Uganda nor Rwanda had significant mineral resources, yet both had become major exporters following the 1996 conflict. Although Uganda had officially withdrawn its army from DRC a few years earlier, its proxy fighters were still smuggling contraband minerals across the border. Nkunda's militia did the same for Rwanda. As Béatrice Le Fraper du Hellen and I had discussed during our meeting in The Hague, this instability seemed to suit the mining companies operating in DRC, many of them Canadian, just fine. They were happy to finance the latest coup if the new leader seemed likely to offer them a better deal. Over dinner, Valentin told me he had contacts in Gisenye, the Rwandan border town nearest Goma, so I agreed we should make the journey together.

The two-lane highway between Kigali and Gisenye, built by the

Chinese sometime after Rwandan independence, follows a precarious route through the mountains. An engineering triumph in its time, making life much easier for farmers who wanted to sell their products in the capital or get essential supplies quickly, the narrow paved road is now crumbling at the edges and full of treacherous potholes. As testimony to the high price paid by foreign road workers who died in accidents or from tropical diseases, there is a Chinese cemetery near the summit. Buses and huge trucks laboured uphill, belching noxious diesel fumes, alongside an array of peasants pushing carts and bicycles laden with impossible loads.

Valentin fell asleep in the seat next to me, his head lolling back and forth as the bus careered around sharp curves, pedestrians and oncoming traffic. We had a busy two days in store. Valentin was confident he could arrange interviews for me with people affected by the genocide, as well as refugees from the continuing violence in the DRC. According to the latest report, Nkunda's troops were dug in a few miles north of Goma as MONUC, the United Nations peacekeeping force, stood by, unable to use its arms unless fired upon. Citizens of Goma, enraged by the odds stacked against them, had taken to throwing rocks and insults at UN troops.

On my left, bulldozers and Caterpillars were busy dismantling one side of a mountain for its ready supply of sand, no doubt to be used in road and house construction. So much for the "Land of a Thousand Hills"; from now on it would be 999, and subtracting. On the remaining portion of the mountain, in a scene worthy of Brueghel, farmers continued working their patches of ground, growing and cultivating as if everything were normal. Along the section of highway being widened outside Gisenye, countless houses and small businesses had been marked for demolition with a huge red X, like the doomed houses and temples I'd seen in the soon-to-be-flooded gorges of the Yangtze River Valley. Suddenly, our bus veered into the oncoming lane to dodge a pothole, heading directly towards two young girls and a man on a bicycle that wobbled precariously under a long sack of carrots. At the last moment, the driver swerved back into the right-hand lane, but not in time to save the carrots, which disappeared into the bushes beside the road. During the *manoeuvre*, a young mother with three kids seated in front of me lost control of the bag of hard-boiled eggs she had purchased at the last stop. We were packed in so tightly it was impossible to bend over as the missing eggs rolled back and forth on the floor. Beside the road, in the shade of several spindly eucalyptus trees, their leaves turned down to reduce exposure to sunlight, two boys tended a few goats and had a go at their makeshift football.

Other than its location on the shores of Lake Kivu and the ring of surrounding mountains, Gisenye had little to recommend it. Poverty, neglect and volcanic action had left their mark everywhere in the so-called business section, which was laid out on a grid but had the appearance of a ghost town in the making. Human efforts had failed to keep up with the ravages of time. The broad, barely passable, lava-encrusted streets were unpaved and corrugated by runoff from the hills. Houses and shops, lacking foundations, had settled at odd angles into the earth, reminding me of an old Yukon gold-rush settlement.

I spent several hours with Gilbert, a long-time friend of Valentin, and his family, who lived in an elegant old house that had seen better days. Poverty and war had reduced this household, like the rest of Gisenye, to near penury. Almost cadaverously thin, Gilbert had a beatific face that shone with intelligence. As an artist and young man, he had studied in Germany, and he spoke several languages. As English was not one of them, Valentin was obliged to translate our conversation from Kinyarwanda into French, which severely limited my comprehension. He delivered his thoughts with humour and authority, as if he were instructing us on the right way to conduct our lives. I recorded the interview, which I hoped to examine back home at my leisure, dictionary in hand, but somehow managed to lose the micro-tape during my travels. I remember, however, one very intense moment when Gilbert was explaining how his family had once been successful farmers, before the government began confiscating lands, taking whatever was arable or had a good view of Lake Kivu and offering it for sale to the new bourgeoisie. Now, they barely survived. Gilbert's son, Thierry, who was studying fine art at school but had little hope of making a living by pursuing his inherited talent, was keen to show me his creations.

As we talked, various members of the extended family—a brother-in-law, perhaps, with a wife and a small child, each as gaunt as Gilbert and Thierry—drifted into the yard and stood listening to the conversation. When the interview, such as it was, ended, Valentin rounded the family up for a photograph. I had arrived with nothing to offer them, not even a small gift of food, but promised to send via Valentin copies of the group photos I took. In one of those photos, sunlight reflects off the right side of Gilbert's head, leaving the rest of his serene face in shadow. The tiny child in his blue-and-white striped T-shirt stares into the camera with a look of wonder, while his father, in a discoloured white T-shirt on which I can make out the date July 1997 and the words "Corunna" and "Celebration," looks on with a pained, sober expression; he has nothing whatsoever to celebrate.

Valentin's other promised interviews did not materialize, so we spent the late afternoon and evening at a quaint old colonial B&B by the lakeshore with his teenaged son Luka, who had crossed the border from Goma to visit. I'd decided that the rebel attacks sweeping across North and South Kivu were reason enough to stay in Rwanda. However, being only several hundred yards from the border and hearing Luka's comments about the ease of crossing, I could not resist the temptation to check out the situation in Goma for myself. Valentin had left his refugee documents in Kigali, so I paid the thirty-five-dollar visa fee and slipped across the border in the company of Luka without incident.

We were whisked into town on the backs of two motorcycles. I had expected a huge military presence in Goma, even on a Sunday morning, but had not anticipated the different sort of hellhole that awaited us. In 2002, the step-sided Nyiragongo volcano had erupted, sending a six-foot surge of molten lava racing through the centre of town at a speed of sixty miles per hour, wiping out the cathedral, houses and businesses—4,500 buildings in all—leaving a nightmarish landscape of jagged, charcoal-coloured rock ugly enough to make a moonscape look inviting. It was obvious that the inhabitants were barely hanging on.

While it makes good, rich soil when broken down, lava is anything but friendly in its initial hardened form. Houses and fences had been reconstructed using chunks of the knife-edged rock, but there were no gardens, lawns or spots to recline. According to Luka, shoes and car tires lasted only a couple of months on this terrain, yet women and children walked barefoot in the streets. A child sat beside a tiny pyramid of potatoes, hoping for a sale. Young men struggled to push their *tricos*, crude wooden scooters with handles the shape of Highland cattle horns, balancing huge loads. Children in rags and flip-flops staggered past lugging plastic containers of water. The rags, I could see, were spotless, all scrubbed and clean for a Sunday morning. Every few blocks, we heard the sound of hymns in makeshift churches constructed from sticks and plastic sheets. One jury-built edifice had been painted orange and bore a sign with the words EGLISE AMENA (Army of the Evangelical Mission of Neo-Apostolics), followed by a phone number. Outside a tiny clapboard lean-to, on which a weathered sign advertised Abdoul Coiffure, a woman bent over a tattered blue plastic container washing clothes, then spread them out to dry on the volcanic rock. Farther along, where a water pipe had burst in the middle of the street, several enterprising lads had set up an impromptu truck wash.

As we passed a Red Cross building with two NO GUNS signs out front, a convoy of white UN vehicles appeared down the street. A pickup truck rolled past carrying four heavily armed Indian soldiers in regulation blue helmets. Bringing up the rear was an armoured vehicle with two soldiers sitting on top behind a mounted machine gun. The driver's helmeted head poked incongruously from the open hatch in front as he waved an index finger back and forth to indicate photos were not allowed. Goma, however poor and devastated, was rich in contradictions. Several beautifully dressed families moved through the rubble on their way home from church; a flatbed truck laden with impoverished labourers and IDPs crossed an intersection in front of a billboard that advertised Mona Lisa cosmetics; a town monument painted orange and yellow bore the words PARTAGER NOTRE MONDE, which translated as "share our world," though *partager* could also be interpreted as "split" or "divide," exactly what the rebels, corporations and the DRC's opportunistic neighbours wanted.

Valentin nodded off again on our return trip to Kigali, content that he had fulfilled his part of the mission. I was satisfied, too. I'd had a helpful glimpse of what I could expect when I returned to DRC early in the new year. For the time being, there was a flight to catch and contacts to meet in Uganda. My mind drifted back to a conversation a few days earlier with Simon Ntare during my visit to WE-ACTX, Women's Equity in Access to Care and Treatment, a medical centre catering to survivors of rape who had contracted HIV during the genocide. These women and their infected mates and offspring were being treated with antiretroviral drugs and given counselling to help them cope with the shame and poverty that often came with their condition.

"I never thought I'd end up running an AIDS centre," confided Simon, who had worked in Tanzania but returned to Rwanda after the genocide to offer his services as a director of customs for the new government.

WE-ACTX, with considerable funding from the U.S., cared for five thousand patients. The waiting room was packed every morning, Simon told me, with some people sitting on the floor. In addition to medical and psychological support, the clinic also operated a craft centre where the women sewed handbags, computer sleeves, backpacks and other items to raise funds. As we picked at a light lunch, Simon shared his concern about the lack of doctors and trained nurses, the inadequate supplies. When the talk shifted to questions of post-genocide justice, he surprised me by suggesting that perpetrators might have a harder time coping than victims, since grief can be easier to overcome than shame.

Then, out of the blue, he made a comment that put his own work into a slightly altered context.

"The problem here is capitalism," he said. "It's making people greedy and less caring." In the rush to rebuild the economy and establish itself as a modern, reliable state, he explained, Rwanda was in danger of forgetting the huge gap between rich and poor.

I did not know what to make of the things I'd witnessed, the stories people had shared with me, the scale of the violence. American author and activist Terry Tempest Williams, who spent some chastening days in post-genocide Rwanda, had concluded, "This is a hell of our own making—those who killed and those who looked away." From what I had seen, I had to agree with historian Martin Meredith that the genocide "was caused not by ancient ethnic antagonism but by a fanatical elite engaged in a modern struggle for power and wealth using ethnic antagonism as their principal weapon." The carnage and human suffering that followed the breakup of Yugoslavia had a similar pattern; according to former Canadian ambassador James Bissett, ethnic groups lived together peacefully there until they were used as "pawns in the game of big power politics." The same card was apparently being played in DR Congo, with minerals rather than power as the jackpot. Unquestionably, racism did lie at the heart of the West's refusal to intervene militarily in the Rwandan genocide, in our failure to demand that the United Nations send in reinforcements or that NATO intervene directly. France alone sent in troops, but for the wrong reasons—vested interests in support of nations of the Francophonie—and only in the final stages of the genocide, when the slaughter was mostly complete. Opération Turquoise served primarily to provide safe passage for the overthrown Hutu government and its butchers into neighbouring countries. The resulting refugee crises would finally awaken the guilt of the international community and produce a colossal outpouring of money and aid—in support of the perpetrators of the genocide whose crimes had been so reprehensibly ignored.

I'd been reading a copy of Sven Lindqvist's *"Exterminate All the Brutes"* I'd picked up during my stopover at Entebbe airport. Lindqvist's title, drawn from Conrad's *Heart of Darkness*, is the phrase offered as Mr. Kurtz's final solution to the "Africa problem." Lindqvist's daunting task is to locate the origins of this sentiment in European thought, and he makes it abundantly clear that racism is an essential ingredient of imperialist ideology. To justify the exploitation or murder of the Other, to assert your right to take over his property or territory or to consciously ignore atrocities, especially genocide, you must believe that you are dealing with people of lower or inferior races. In the case of Africa, as Lindqvist puts it, "Too many Europeans interpreted military superiority as intellectual and even biological superiority." His brilliant but gruelling analysis of the origins of a rhetoric of extermination comes to a number of disturbing conclusions, including this one: "The pressure of the hungry and desperate billions has not yet become so great that the world leaders see Kurtz's solution as the only human, the only possible, the fundamentally sound one. But that day is not far off. I see it coming. That is why I read history." I had to wonder if—through political and economic interference on the one hand and deliberate neglect on the other—extermination was not already high on the New World agenda.

On the road ahead, I could pick out an overturned cab and loaded flatbed trailer exactly where I'd seen them the previous day. Things move slowly here, I thought. And progress, however it might be defined, is likely to be almost imperceptible, achieved at the micro rather than the macro level. As the bus to Kigali climbed higher and higher amidst racing clouds and the ghosts of dead Chinese labourers, I felt fingers moving across my shoulders, followed by soft laughter.

Two young Rwandan women in the seat behind had done me the favour of brushing dandruff off my shirt.

four *They Kill Us for Their Sport*

WE WERE gathered, just the three of us, at a compound outside Gulu in northern Uganda, where Evelyn was speaking the unspeakable. We had arranged some chairs next to a small building to take advantage of the shade. Evelyn perched on the extended concrete lip of the foundation, a short distance from Victor and me, a plastic cup at her side. She wore a blue-and-white patterned blouse with columns of tiny red roses and a green skirt with larger yellow and blue flowers, garments completely at odds with the sad, lost gaze in her eyes. Her face was slightly disfigured, but most of the damage she'd suffered was not visible. She gave the impression that it was all she could do to take another breath, to carry on living. She answered my questions with such reluctance, such apparent agony, while Victor translated, that I found it difficult to concentrate.

"Mostly, I'm silent now. Back home, the cleansing rituals did not work. Mine was a leper's welcome. My husband can't stop thinking about the older men I was given to, Joseph Kony's people. I was abducted three times, age seven to fourteen. Kill or be killed, they said. Mother, Father, forgive me." I looked up from my notebook at Victor, the significance of Evelyn's cryptic remark spreading through me. He shook his head and continued interpreting.

"The village was on a small river by the Sudan-Uganda border. The third time, a rebel appeared in the doorway. He'd used a rope to pull himself across during the flood. First I carried his looted property, then his child."

Nothing is simple, nothing smooth, only jagged like the bottle cap Evelyn picked up from the ground and shifted from hand to hand, the sharp points digging into her flesh. "Of course, I'm bitter. Who will defend me? Who can explain it all?" She turned away, knocking over the plastic cup.

Two motorcycles passed on the road, the whine of their small engines cutting through the silence that engulfed us as we sat, motionless, in the sweltering shade.

MY PLANE SWOOPED low over Lake Victoria and I arrived at Entebbe International Airport for a second time, prepared for neither the illusory lightness that awaited me in Kampala nor the deep reservoir of pain that would draw me farther north. A few minutes after leaving the airport, just beyond the UN installation containing several planes and fifty gleaming white SUVs waiting to be dispersed to the new troops destined for DR Congo, my taxi coasted into a small gas station, where the driver purchased half a gallon. The tiny infusion did not even budge the needle on the gauge. The car inched again into the heavy Saturday to-market traffic and the choking cloud of diesel fumes for the twenty-mile trip to the Ugandan capital. English rules of the road applied here, along with the death-wish driving habits of the French. Whenever traffic slowed to a stop, the driver turned off the engine.

As we continued along the Jinja Road towards Kampala, driving through the vibrant commercial chaos of ordinary life—a plethora of food stalls, tire stores, cellphone dealers, undertakers, pharmacies and metalworking shops—I sensed a sea change taking place in my body. Although I was excited by the bustle and the ridiculous billboards advertising Nile beer as "the true sign of progress," my breathing grew deeper and I could feel myself succumbing to the need for sleep. Only the constant honking and the bumper-to-bumper jockeying of cars, bicycles, motorcycles, trucks and buses kept me awake. I'd left behind the compulsive orderliness and the impenetrable cloud of suffering that hung over Rwanda and was beginning to feel at home. I checked into the New City Annex Hotel on Dewinton Street, situated next to a wonderfully grotesque plaster replica of an ox head mounted on a globe, then located a nearby ATM that accepted my debit card and delivered a hefty wad of the local currency. I could have kissed the ground in gratitude.

The strangeness of this reaction did not escape me. Idi Amin's reign of terror in Uganda, followed by Milton Obote's tumultuous second term in office, the insurgency of the National Resistance Movement that brought Yoweri Museveni to power and kept him there and the ongoing murders, mutilations and abductions perpetrated by Joseph Kony's Lord's Resistance Army in the north should have been sufficient cause for sleeping lightly with prayer beads in hand, a gun under the pillow and a double lock on the door. Instead, I slept as if drugged. Strolling through the streets after my power nap, I revelled in the relaxed atmosphere of Kampala. Everyone spoke a modicum of English. An invitation to dinner that evening also helped dispel the usual first-day jitters.

My hosts, Cathy Watson and William Pike, lived in a gated compound in an elegantly sprawling house with exquisite gardens and two rental cottages in the suburbs of Kampala. They were highly regarded old Africa hands who had come to Uganda in the 1980s. William had interviewed Museveni in the bush, he told me over drinks, convinced the rebel leader would soon take control of the country. He imitated the raised eyebrows of his interrogators when he was questioned later by British intelligence. William was born in Africa, but educated at Marlborough College in England. Cathy was Australian, but she had trained as a nurse in New England. The two of them met in a Swahili language class in London, and by the time they arrived in Uganda, where William founded and edited the *New Vision* newspaper, Cathy was also working as a journalist.

We were seated in an alcove on the terrace overlooking the gardens as the sun began to set, casting an orange glow on the outer buildings. A car moved slowly up the driveway. While William went to greet the new arrivals, Cathy described covering a rebel advance in Uganda. Officers of the government forces had to beat their troops with sticks to get them to fight, a hopeless situation as the men were tired, hungry and afraid and the rebels were rapidly advancing. Running alongside the retreating government soldiers, but afraid of being shot in the back, Cathy dropped into a trench where she could hear feet squelching past her in the wet ground. A government soldier helped her wade shoulder-deep through a swamp, his gun and her camera and notebook held high overhead. They had the luck to intercept a replacement unit and were escorted to safety.

Not long before my arrival in Uganda, William had been encouraged to resign as editor of *New Vision* after running articles critical of the government and Museveni, but he still owned and operated Capital Radio in Kampala and was commuting regularly to Nairobi, where he had started another newspaper that doubled as the printer for a variety of weeklies. When he advised the editor of one of these dodgy political tracts that it might be safer to advertise his publication as "a moderate voice," the man had protested: "No, then the extremists would kill me and my family." Cathy had initiated a tree-planting program in Uganda and founded a controversial organization called Straight Talk that provided sex education for young people in a climate where AIDS was still rampant and women's rights were largely ignored.

As the sun disappeared, mosquitoes crashed the party. Three other couples had joined us for dinner, so we retreated to the dining room. Delicious smells emerged from the kitchen, where the chef was busy. John, still bearing scars from the rebellion in his Ethiopian homeland, ran a printing business, which produced the *Monitor*, an opposition daily. He invested in gold exploration as a sideline. His wife, whose name I did not catch, was in business, too, importing mostly from Dow Chemicals, she told us. Filipo and Luciana, two Italian doctors, had been in Uganda since 1980. They'd spent ten years practising in rural areas, where Filipo had been captured by rebels but released after only a few hours because one of the leaders recognized him. "You treated me in the hospital," the man had explained. The village women had taken Luciana to task more than once for letting her husband carry groceries home from the market, and two of the local nuns always giggled when they saw her and Filipo walking to work with a female friend employed at the hospital. She asked the Mother Superior what was so funny.

"They think the doctor has two wives, but only loves one of them because the other is so small and skinny."

A delicious pasta dish was served in a large wooden bowl along with a salad covered in fresh cilantro and slices of mango. William opened a second bottle of wine and topped up our glasses. Conversation moved easily back and forth across the table, words washing up against the simple, elegant decor, a carving, a striking tapestry.

"Speaking of wives, Musheshi, what is your opinion of polygamy?" Our host was addressing his Ugandan friend, a businessman and educator who'd helped set up a technical training institute for women.

Musheshi launched into a detailed argument for polygamy, suggesting the main reasons for it were economic and political, to provide additional labour in the family unit and extend influence in the community.

"Not just for sexual variety?" I couldn't resist asking. "More mouths require more food," I added in an effort to redeem myself.

Musheshi worked his facial muscles into an indulgent expression.

"My question had nothing to do with morality," William explained. Two of his employees had grown up in polygamous families, and both were having difficulty adjusting to life in the city. One, polygamous himself, was debt ridden, his life in tatters. The other, as the child of a less-favoured wife, had never learned how to communicate his frustrations; on the job, he simply stored them all up until he could no longer cope and wanted to quit.

Musheshi was not convinced. "It has everything to do with the personality of the individual father, regardless of whether he has one wife or many."

The meal progressed to dessert, a fruit pastry, and coffee, and the discussion shifted to politics and the future of Museveni, who had legislated for two terms in office when he came to power in 1986, then changed the rules. These days, it was beginning to look as if, like Zimbabwe's Robert Mugabe, Museveni might consider himself entitled to a life term. I'd read Museveni's 1998 book *Sowing the Mustard Seed: The Struggle for Freedom and Democracy in Uganda*, a sanitized but revealing account of the revolution. The book detailed the rigorous code of conduct Museveni demanded (but did not always get) from his rebel soldiers in the National Resistance Army (NRA) or from fighters inherited from the Uganda People's Defence Force (UPDF); his efforts to elevate the status of women; and his establishment of a human rights commission. However, there was an oddly colonial ring to his comments about the backwardness of his people and the need to bring them "civilization," mostly by developing private enterprise. And nowhere had he adequately addressed the shortcomings of his regime, particularly its growing corruption.

William, a long-time friend and supporter of the president, nonetheless felt Museveni should step down and assume the role of wise elder statesman for the African continent. Filipo disagreed, convinced that Museveni would not be replaced by anyone half as reliable, with disastrous consequences for the country. The question most troubling me was why the human rights situation in the north had never been seriously addressed, why poverty there was rampant and Kony's ragtag band of guerrillas were permitted to kill, mutilate and terrorize the Acholi inhabitants of Kitgum and Gulu. But the other guests were already rising from their chairs.

The next morning, I rose early. I phoned my list of contacts, all of whom were either not in the office or unavailable. Disappointed, I decided to check out downtown Kampala, starting with a shoeshine by a young entrepreneur in a blue T-shirt at the corner of Dewinton and Siad Barre, a street named after the dictator whose ousting in 1991 heralded the social and political collapse of Somalia. Replacing the tainted dust of Rwanda with a fresh layer of shoe polish seemed a positive gesture. I sat on a stool, my stocking feet on a small scrap of carpet, and watched the traffic as layer after layer of the brown cream was applied to my shoes. A woman wearing a white muumuu with large black spots stepped into the street between a parked black sedan and a blue pickup, forcing a Jeep to stop, then paused under a Crane Bank advertisement to adjust the load of bananas she was balancing on her head. Sunlight was already intense enough to mute the trees and red tile roofs on the adjacent hillside. And across the street, various negotiations were under way, for newspapers, the ubiquitous phone cards and new pants, thanks to a street merchant outside the New City Annex carrying half a dozen pairs, one of them draped over his arm for display.

I tried on my shoes, which were now transformed and gleaming. The pint-sized maestro accepted my over-large payment as if it happened every day, tucked the bills in the back pocket of his jeans and gave his two friends or partners a high-five sign. Freshly shod, I dropped into the National Theatre nearby to purchase a ticket for a performance the next day, then made my way circuitously to the Speke Hotel, where I planned to read the paper, check my e-mail and, later on, meet with Jackie, the sister of my Ugandan journalist friend Sam, whom I had met in Kigali.

Kampala is built around a series of hills and bursts at the seams with parks, churches, bookstores, attractive high-rises and appalling slums, traffic congestion, nightclubs, petrol shortages, boutiques, bistros and beggars. In short, it has all the contradictions of a modern city. Even the Independence Monument, with its sculpture of mother and child triumphant and a wall of revolutionary murals, conveys mixed messages, juxtaposing scenes of a Jeep full of armed men, a strategic bush conference, a teacher with a class of enthusiastic students on benches and a victorious Museveni sporting a suit worthy of a real estate salesman and holding the model of a red-roofed house, a symbol of the splendid future in store. The president had delivered on many of those promises, though not for the young woman cradling an infant in her arms, both of them asleep on the sidewalk as I passed. There was even a warning sign in red ink beside the independence sculpture that said IDLERS ARE NOT ALLOWED AROUND THIS MONUMENT, an irony I knew Sam would appreciate.

Although some streets were lined with daffodils and chrysanthemums and signs of affluence were evident, they had not touched the life of the legless beggar I encountered outside the Speke Hotel. "My name is David," he said, looking up at me from the stretch of pavement that served as a traffic divider. A parade was passing by, with a military band in green berets and several columns of young people in yellow T-shirts carrying a banner indicating they were the children of fallen comrades in Uganda's National Resistance Army, marching to a fund-raising dinner. As we couldn't talk, I shook David's hand, gave him a thousand-shilling note, then slipped into the hotel, where I purchased a newspaper and took a seat in the streetside café.

According to a small announcement on the front page of *New Vision*, Joseph Kony, whose rebel force was currently ensconced over the border in the Garamba National Park in DR Congo, was expected to sign a peace accord bringing to an end two decades of violence. The signing, a few days hence, would take place in the Sudanese city of Ri-Kwangba, two hours south of Khartoum by air. Kony's failure to show up at previous peace initiatives no doubt explained why the announcement took second place to a more sensational story, POLICE SEARCH FOR BOY'S HEAD, the latest update on the ritual murder and beheading of a twelve-year-old boy, allegedly instigated by a city tycoon and carried out by an obliging witch doctor and his wife. The boy's genitals, too, were missing.

A guerrilla force is not easily defeated by conventional troops, as evidenced by the success of the Vietcong, the Khmer Rouge, the IRA, and the Taliban. Guerrillas can disappear into the landscape or blend in with the local population. However, this alone could not explain the two-decade survival of Kony's movement. Although often dismissed as a madman and a psychopath, Kony showed his tactical genius when it came to deploying his small, disparate and highly mobile bands of followers. Funds provided by international supporters, including the government of Sudan, were not squandered on luxuries or squirrelled away in foreign bank accounts, but instead invested in the latest satellite phones and solar battery chargers to allow constant, reliable contact. Only occasionally would international media attention stimulate Ugandan efforts to halt the ongoing atrocities of the LRA. One such event was the abduction of 139 schoolgirls from St. Mary's College, Aboke, on October 10, 1996. The plight of the Aboke girls and Sister Rachelle's relentless struggles on behalf of her missing secondary students brought Pope Paul II, the UN's Kofi Annan and other international leaders into the picture, forcing Museveni, for a time at least, to assume a more active role in confronting the LRA.

In the same issue of *New Vision*, I learned that a humanitarian crisis was brewing as more and more refugees poured into Uganda from the eastern regions of DR Congo, the result of violence from both Congolese militias and the LRA. A Ugandan member of Parliament was reported to have been involved in a traffic accident en route to the funeral of a colleague who had died in a similar accident the previous week. Although four ambushes had occurred over the previous month in the country's Karamoja region, causing three deaths and thirteen injuries, a UN report cited in the paper credited increased security as reason to hope for better times. Lord Steel of Aikwood, a former member of the House of Commons and the Speaker of the Scottish Parliament, who was in Uganda for talks, had told an interviewer that he admired Museveni and understood the need for unity, but disagreed with him about delaying a return to multi-party politics in Uganda. The newspaper centrefold contained a fawning article about Congo warlord Laurent Nkunda, in which the reporter recorded his own struggles going to and returning from the rebel camp but had nothing whatsoever to say of the general's warped vision and brutal methods. A comment by the ethics minister had been singled out as Quote of the Week: "They [homosexuals] say they are being harassed and discriminated against. This is not true. No one should discriminate against a sick person. They are entitled to getting drugs like all Ugandans."

With a half-hour left to wait for Jackie, and having had my fill of uplifting stories, I took a short stroll through the spacious Sheraton Gardens, replete with palm trees, bird of paradise, red heliconia and other extravagant tropical vegetation, then made my way along Nile Road past a monument erected for the recent Commonwealth Heads of Government Meeting. As I admired the family of sculptured figures holding up a Ugandan flag and looking skyward to an uncertain future, I recalled a related item in *New Vision*, a few pages after the decapitation story, about inflated contracts stemming from that important meeting and a photograph of the queen during her official visit to Uganda, her own head still very much intact—and sporting a large pink hat.

Jackie arrived at the outdoor restaurant at the Speke Hotel just as I got back. A student of literature at the university, she was very much attuned to the political realities of Uganda. After some small talk about graduate studies and her brother, Jackie presented me with a copy of the Kigali *New Times*, in which Sam had published a more extensive article about my visit to Rwanda.

"Sam was a big help to me in Kigali," I said.

Jackie nodded her head in agreement. "Yes, I can imagine. My brother is a dreamer and an enthusiast, but I love him dearly. I miss him, too."

We talked about books and Jackie recommended several local authors, including the late Okot p'Bitek; novelist Austin Bukenya who, in *The People's Bachelor*, pillories the paper socialism of university elites; and Arthur Gakwandi, whose novel *Kosiya Kifefe* is an exposé of intrigue and corruption in post-colonial Africa. One of her cultural icons, Andrew Mwenda, the fiery journalist and founder of the *Independent*, had been jailed briefly for an article he wrote about the government and Museveni, Jackie told me. He had used his cellphone to file regular reports and interviews with other detainees from inside the prison. Charges against Mwenda were dropped, an indicator of both the confidence of the ruling elite and the growing maturity of the body politic. After a conversation about the recent land scandal in which a government minister was implicated, I asked Jackie her thoughts on the coming elections in 2011. She answered without hesitation.

"I've lost faith in politicians because of the speed with which they abandon lofty ideals for the advantages and privileges of power."

The night before, in my hotel room, I'd dipped into the pages of a book called *Uganda's Revolution 1979–1986: How I Saw It* by former rebel, bush-fighter and NRA commander Pecos Kutesa. Kutesa was modest and disarmingly candid in his description of the failures, triumphs and setbacks of the revolution, acknowledging that freedom of assembly had been difficult to achieve and that while Uganda was no longer "the pauperized laughing-stock of the world," much remained undone:

> The building of an independent and self-sustained economy, which was one of the goals of Uganda's revolution, has so far remained an elusive dream . . . Instead Uganda has become more and more dependent on donor largesse and budget support, making us vulnerable to external pressures and dictation. The architects of Uganda's revolution did not anticipate this turn of events. The promised modernization of agriculture is yet to be achieved and the economic gains of the revolution have not evenly trickled down to the peasantry, which is the backbone of the NRM regime. Many of our people, especially in northern Uganda, remain trapped in abject poverty and backwardness, which means that the mission of Uganda's revolution was not fully accomplished.

His having chosen to remain a soldier rather than pursue the more powerful and lucrative career as a politician that, as a hero of the revolution, would have been his for the asking gave Kutesa's assessment considerable weight.

I was keen to continue my conversation with this bright, informed young woman, who spoke with a healthy dose of cynicism, but one of her professors emerged from the hotel lobby, where he had been making travel arrangements for a trip to the U.S. to marry his fiancée in DeKalb, Illinois. After the introductions, he said to Jackie in a half-joking voice, "You should be studying."

"She's been working overtime," I assured him, "trying to instruct me in literature, journalism, history and the finer points of rhetoric."

"Jackie is a brilliant student," he replied, excusing himself. Jackie quickly took her professor's advice and headed off to the university.

Walking back to my hotel, I noticed huge blotches of white paint dotting the sidewalks of Kampala, which struck me as incongruous in a city anxious to put on a good face for the world. As I stopped to examine one of these splotches, a cacophony of shrieks erupted overhead. I dashed a safe distance along the sidewalk, then looked up, but not before a measure of this white substance splashed on my newly polished shoe. A huge vulture-like creature was alighting in the branches above me, and her ugly offspring had turned up the volume in anticipation of a feeding frenzy. Uganda has more than a hundred exotic and endangered birds, I read later in my Bradt guide, but this was not one of them. The marabou storks, known locally as *kalooli*, are neither elegant nor exotic and have colonized downtown Kampala, nesting in every sizable tree and laying down carpets of excrement on the sidewalks. The bird is described in the Bradt guide as "a macabre carrion-eating stork, 1.5 m tall, with a large expandable air-sac below the neck, and with a black-and-white feather pattern reminiscent of an undertaker's suit." The description would seem even more apt when I learned the next day that marabou storks often hover above the placenta pit at nearby Mengo Hospital.

MENGO HOSPITAL'S SPRAWLING facility occupies a hillside compound that served as the capital for the Buganda ruler Kabaka Mwanga and subsequent *kabakas*. A short distance north of Kampala, and dominating the surrounding suburb, it's the oldest hospital in East Africa, founded in 1897 by Albert Cook of the Church Missionary Society. In 1958, it became a private institution. Mengo's first operating theatre, a mud hut with thatched roof, was expanded into a twelve-bed structure that was destroyed by lightning in 1900. Four years later, the main hospital was built in colonial style, replete with shuttered windows, wooden beams, brick and plaster walls, cement floors and a corrugated metal roof. It was one of these original wards, now reduced to the status of a health hazard, that the Friends of Mengo Canada had agreed to renovate. Volunteer Bill Gilchrist had arrived in Uganda a few weeks earlier to help supervise the renovations. His wife, Cathy, a nurse, was training local nurses in the techniques of therapeutic touch.

Bill invited me to meet him on the job. After he introduced me to the crew of volunteers and local workmen and explained the extensive renovations under way—which included replacing rotten beams, stripping and replacing plaster walls, adding solar panels and installing new showers for staff and patients—we set out on a tour of the facilities, starting with the dental unit that had been completed the previous year with foreign funds and was now the pride of the institution. It had both private and public sections, labs, a backup generator and a dozen high-tech reclining chairs. Dental care is not free in Uganda, though, and we saw only three patients in the clinic that afternoon.

This was Bill's third stint as a construction volunteer at Mengo Hospital, and he was sounding a bit jaded. Permanent screens had been installed in the renovated ward because nurses were in the habit of leaving windows open well into the evening, inviting in hordes of mosquitoes. As a result, patients arriving with pneumonia might well leave the hospital having contracted malaria. The wards were hopelessly understaffed; nurses had to do the cleaning when they arrived each morning before attending to the needs of patients. In the maternity unit, mothers and their newborns were often left on the floor for hours amidst cleaning equipment and debris. In the previous two weeks, six new mothers had died because of bad hygiene and poor medical care. Cathy was beside herself, Bill told me. In her thirty years as a nurse in Canada, she'd seen only one such postnatal death.

We toured the Eye Department, then the HIV/AIDS Day Clinic, a delightful open-air facility, greeting staff as we went. When I asked Jim Sparling, a long-time support doctor, about the effectiveness of foreign aid in Uganda, he launched into an anecdote about being asked to diagnose a staffer working for the U.S. Agency for International Development who was not seriously ill but was nevertheless being sent home by Learjet. USAID had invited Jim to attend the individual's going-away party for five hundred dollars a plate. Gourmet food and booze were being flown in from Paris to the organization's palatial mansion with its marble fixtures, king-size beds and wall-to-wall servants. Disgusted, Jim had declined.

After hearing about the shortage of supplies at the hospital, especially antiretroviral drugs, about obstructionist administrators and plain incompetence, Bill and I retired to the local watering hole for a beer, near the main intersection in Mengo where, he informed me, three people had been killed and hundreds intimidated during the previous elections. Our plan was to meet Ray Wood, a British doctor who was a legend at Mengo Hospital. Ray was already installed at the bar, holding forth. A gruff character who took his job seriously and worked hard but was dismissive of his own contributions, Ray hated the bureaucracy.

"Accountants? Bloody wankers! Those assholes up there," Ray was on a roll, his finger pointing at a bottle of Scotch on the shelf behind the bar, but meant to indicate the administration offices on the hill, "they don't care about patients. They're only interested in cheating me." I presumed he meant chiselling away at his budget.

Having seen for myself the good work being done by the hospital's dedicated staff under very challenging conditions, I decided to push my new companions a little. I mentioned having told an acquaintance of my planned visit to Mengo Hospital; he had rolled his eyes and said: "The conventional wisdom in Kampala? 'Go to Mengo and die.'"

The remark prompted a raft of objections and denials. Clearly, my drinking pals loved this dirty, ragged, inefficient institution.

The following day, I paid a visit to Mulago Hospital, a slightly upscale medical institution in Kampala, where Juliet Tembe was based with The AIDS Support Organization (TASO). In spite of vigorous advertising and the availability of antiretroviral drugs, Juliet told me, a degree of complacency had settled in, especially among the twenty-four- to thirty-five-year-olds, and the number of new AIDS cases was no longer declining. Although TASO's mottos include "the right to adequate medical care and respect" and "the right to live positively and die with dignity," it is difficult to get this message across in a culture of illiteracy, in which the rights of women are only slowly being recognized and 65 percent of women cannot read. Juliet, who studied English language, literature and education at Makerere University in Kampala, had just completed a PhD at the University of British Columbia in basic linguistics and language preservation. Her aim was to ensure that all Ugandan women were instructed to read in their mother tongue. I recounted a story from Mark Abley's *Spoken Here: Travels among Threatened Languages*, in which he reports the linguists' prediction that 90 percent of the world's six thousand languages could be extinct within a century. Early in the nineteenth century, during an extended visit to Latin America, German explorer Alexander von Humboldt heard a parrot speaking in a Maypures village near the Orinoco River in Venezuela. When he inquired what the bird was saying, the villagers told him the bird was speaking Atures, a language for which there were no longer any human speakers.

Mulago Hospital, not the facility of choice for foreigners or well-to-do Ugandans, many of whom preferred private clinics, was surrounded by dilapidated shacks, home to unemployed men, children in rags, the gaunt faces of women in doorways. Juliet's hope for adequate health care and language protection in such surroundings was not being aided by government neglect and a plethora of crass English billboards declaring that Coca-Cola promised "deep down satisfaction." So many wealthy Ugandans, politicians included, were living high on the hog in the midst of dire poverty, inadequate education and abysmal health care; it was enough to make any hungry, unemployed Ugandan consider the alternatives: crime or rebellion.

My brief exposure to Uganda's hospitals did nothing to reduce my own medical anxieties. Food had been a major preoccupation since my arrival in Africa. This was a pathetic bourgeois concern, I realized, on a continent where most people drink untreated water and are lucky to have one meal a day, but I had work to do and illness would cost me time. So I was grateful to discover the Masala Chat Indian Restaurant, where I settled down for supper and a large bottle of Nile beer. While I tucked into a delicious vegetable curry, I reviewed my itinerary for the days ahead, which included an interview with Joseph Manoba, recommended to me by my contacts at the International Criminal Court. The Masala Chat was popular, with a lineup forming at the door, so I polished off my beer, paid the bill and sauntered back down Dewinton Street, my target the ox-headed globe I had dubbed Oxymoron. It looked absurdly sinister at night. At the far end of the long, bleak corridor that led to my hotel stood the gesticulating cardboard cut-out of a white man, intended to promote the sale of shoes. Although I'd seen him a couple of dozen times already, the figure always startled me. In this context, he was ridiculous, two-dimensional, devoid of meaning and substance.

JOSEPH MANOBA WELCOMED me the next morning into his small ICC office, tucked away safely and inconspicuously in the bosom of another NGO. A lawyer trained at the University of Kampala, he had not found entry into the legal profession in Uganda easy or congenial, he told me, given the high levels of corruption and cronyism. The Uganda Commission for the ICC at least put Joseph in touch with real people and the possibility of justice. As our meeting got under way, he took pains to explain some of the traditional Ugandan justice systems. The Acholi people of northern Uganda practised Mato Oput, a ritual which involved both parties drinking the bitter root of the *opot* tree mixed with blood and *kwet*, a maize-based fermented beverage, from a common cup, or calabash, as part of the reconciliation ceremony after someone was intentionally or accidentally killed. The traditional method of the Lango people, called Kayo Cuk, to deal with a serious crime such as murder involved paying a debt of eight cows—seven of which went to the victim's family and one of which was slaughtered for a feast of reconciliation—and biting a piece of charcoal from the fire of the ceremonial roast. When I asked how these systems could be reconciled with the ICC's mandate to bring major offenders to justice in a high-profile courtroom, Joseph declined a simple answer. Kony had been indicted by the ICC in 2005 and was using this as an excuse for avoiding peace negotiations. Many Acholi thought the charges should be dropped, Joseph Manoba said. They believed Kony could be dealt with through traditional methods, however bitter a pill that might be for some of the injured parties to swallow.

Our conversation ranged widely over victims' rights, reparations and the protection of staff and witnesses. Someone had broadcast the licence number of an individual involved in ICC work in Gulu, Joseph informed me, and that person had been murdered. Joseph was orchestrating theatre performances in Lira, Gulu and Soroti, where victimization and justice scenarios had been acted out in public. A new script was in preparation. I was curious about the ICC's Trust Fund for Victims, an independent body that administers reparations and assists people with the rebuilding of their families and communities. Kony had been in business for two decades, and many of his victims would either be dead or have endured ruined lives by the time he was brought to trial and convicted, I suggested. Joseph explained that these reparations, originally intended to follow conviction, gave immediacy and a more human face to a remote process. I should talk to Victor Ochen, he urged. Victor worked with the African Youth Initiative Network (AYINET) to support war victims of the conflict in northern Uganda. He would be able to answer my questions in more detail.

I had decided early on that it was too dangerous to go to northern Uganda. Nevertheless, I phoned Victor the next morning, giving him an abbreviated version of my trip so far. The enthusiasm and the light in his voice caught me off guard.

"Gary, you must come to Gulu. We have a lot of patients at the hospital right now getting surgery and attention for their injuries. Kony victims. It's a great opportunity. You must come and talk to them."

"I'll let you know," I said, but I already knew I had to get out of the capital, with its noise, its distractions, its comforts.

I spent the evening at the Speke Hotel catching up on correspondence and letting my wife, Ann, know about the change in plans. On the way back to Dewinton Street, I was accosted by two aggressive prostitutes, one of whom threatened to follow me to my hotel if I did not give her money. I could hear her raucous laughter as I took to my heels. To my surprise, David, too, was working the street, begging well after dark. He looked up at me from the sidewalk, delighted.

"You remembered my name!" He pointed to the frayed cloth wrapped around his stumps. "Pray that someone will buy me a wheelchair."

FROM THE WINDOW of the Eagle Air twenty-seater twin-prop plane, I could see two islets in Lake Victoria, each with half a dozen trees, both too small for human habitation. At this altitude, the islets might have been any rocky outcrop in the straits of Georgia and Juan de Fuca off Washington's and British Columbia's west coast. I was sixteen thousand miles from home, more than halfway around the planet, flying over the site of so much history and romance, so much bloodshed and pain. Opening up below, this major body of water in the famous Rift Valley—though not quite the much-sought source of the Nile, the discovery of which cost David Livingstone his life—had been a staging ground for the slave trade, Arabs carting their hapless victims northward, Europeans and North Americans conducting the same cruel transport to ports on the Atlantic coast. More recently, Lake Victoria had served as another link in the infamous Highway of Death along which long-distance truckers, traders and pilgrims carried the deadly HIV virus.

After a brief stop at Pader to pick up several aid workers, the plane proceeded to Gulu. I could see below me the occasional cluster of circular huts with conical straw roofs in a dry, sparsely treed landscape. A single truck bounced along a dirt road raising a plume of dust. I had expected more vegetation, but the land had not recovered from drought, deforestation and the never-ending demand for charcoal. How had the LRA survived so long on this terrain, where there was no hiding place from the air? Night raids, presumably, and a lack of regular government surveillance. Before the pilot completed his announcement that we would be landing shortly in Gulu, the plane touched down on the runway, decelerated noisily and came to a full stop in front of a small shed, where a single taxi waited. The town was nowhere in sight.

Gulu figures prominently in the history of post-colonial Uganda. It was here that Acholi leader Tito Okello, the Uganda People's Defence Force (UPDF) general whose supporters within the military were to topple the dictatorship of Milton Obote, first met Alice Auma, a fishmonger who claimed to be possessed by the spirit of a dead Italian soldier named Lakwena, whose name she would later assume. Although General Okello refused her offer of spiritual advice, Alice Lakwena went on to become a legend in northern Uganda. When Museveni and his NRA forces broke their power-sharing agreement with Okello's Military Council and marched into Kampala on January 25, 1986, the Acholi people felt betrayed. Lakwena, an Acholi Joan of Arc, abandoned her healing practices and took to the field, organizing the Holy Spirit Mobile Forces (HSMF) in a war against corrupt devils in the south. Her movement gathered considerable momentum, in spite of its bizarre combination of witchcraft and Christian rhetoric. She convinced her followers that rubbing shea butter over their bodies would protect them from bullets. Alice's ragtag militia made significant advances before being defeated, after which she took refuge in neighbouring Kenya, devoting herself once again to prayer and the healing arts. But Lakwena's militant spirit grew restless, it appears, and took up residence in former UPDF soldier Joseph Kony.

The Lord's Resistance Army—dedicated to the overthrow of the Museveni government and the establishment of a regime based on the Ten Commandments—ravaged Kitgum, Gulu and much of the Acholi nation, Kony's own people, whom he was purporting to liberate. Entire villages considered disloyal were razed, their inhabitants raped, mutilated, massacred or abducted. Beheadings and dismemberment were common, along with a host of sadistic practices, such as plucking out eyeballs or forcing survivors to kill and then eat the boiled body parts of their own family members. Over the course of two decades, an estimated ten thousand people from the north were murdered, as many children abducted to become child soldiers or sex slaves, huge numbers mutilated and upwards of a million people driven off their land into IDP camps. The collapse of agriculture, the closure of schools and the destruction of medical services rendered Uganda's northern breadbasket a deadly and unproductive disaster zone.

Gulu, the town and the region, had never recovered from government attacks during the civil war or deliberate neglect in the aftermath. Its plight had inspired an invasion of another kind: NGOs. As my taxi navigated the roundabout to my hotel, I glimpsed signs for at least a dozen organizations, including UNICEF, World Food Programme, Food and Agriculture Organization, Legal Aid Project, United Nations Development Programme, USAID and the United Nations High Commission for Refugees. I'd learned, thanks to Jackie, that in addition to being Uganda's ground zero, Gulu was also the birthplace of Okot p'Bitek, one of the country's most famous writers. With his father a teacher and his mother a singer, p'Bitek seemed destined to become an artist. His studies in education, law and social anthropology took him to universities in Bristol, Aberystwyth and Oxford, after which he taught at Makerere University and was director of Uganda's National Theatre before falling out with the regime of Idi Amin. In exile, p'Bitek taught in Iowa, Kenya, Texas and Nigeria, returning to teach creative writing at Makerere in 1982. His most famous work, *The Song of Lawino*, is an epic poem of complaint in the voice of a neglected Acholi wife whose husband has abandoned the old ways and embraced the superficial values of the colonizers. It's a vibrant, entertaining and deadly serious assault on the colonial mentality.

In an essay called "What Is Culture?" Okot p'Bitek makes a spirited case for indigenous culture, which, when not actually ignored or denied, is often stolen, transported and copied by other countries. After describing the "ruthless and heartless extirpation" of African nations by Europeans, he sketches in the transformed landscape: "Huge, ugly Gothic houses, called churches, sprang up on hills, in forests and on the grasslands of Africa; and men and women were wooed to go there on Sunday to worship a Jew, represented as a corpse dangling on a tree." In these alien structures, boys and girls were brainwashed in the values of the new religion. "Thus, most contrary to the African idea that everybody must marry, in the prime of their youth, have a family (and for a man, the more wives and children he has the better), young intelligent, beautiful and handsome Africans were lured to think that wifelessness, husbandlessness, childlessness, homelessness, were a virtue!"

A week earlier, I had attended a performance at the National Theatre in Kampala, keen to experience the vibrancy of Ugandan culture. In the play, a prisoner convicted of murder is offered the chance to confess to a priest before her execution. Instead, she harangues the priest and converses with the ghosts of former friends and acquaintances, who pop up in the wings from Hell and Limbo to argue from their vantage points. As these ghosts are invisible to the priest, he thinks all of her comments are addressed to him, which makes for one or two amusing moments. On one level, it was a play about forgiveness, letting go of anger, particularly the self-loathing kind. However, there was not a scrap of social commentary, no analysis of post-colonial social and political conditions and nothing, in terms of music or philosophy, that might have been viewed as indigenous. When the curtain dropped, the director appeared on stage and invited anyone in the audience who wanted to talk about their spiritual problems to stay behind. Unlike Bertolt Brecht's Theatre of Alienation, which actively engages audiences with a view to effecting social change, this seemed to me a cheap morality play, only slightly less offensive than evangelist Billy Graham's fire-and-brimstone altar calls. Here in Okot p'Bitek's hometown, I wondered what he would have thought of the play and the performance. At least he would have approved of the ironic treatment of the priest and the caricature of the Catholic afterlife.

I called Victor from the hotel. His voice still radiated light. "I saw your plane come in for a landing at the airport twenty minutes ago and told Richard: 'Gary will be on that plane.' Come over to the hospital as soon as you can."

Before I had a chance to unpack my bags, I was being whisked away on the back of a *boda-boda* along the bumpy dirt road to St. Mary's Lacor (pronounced la-chor) Hospital two miles south of town. Cyclists and pedestrians, among them several priests, competed with motorcycles and the occasional truck or SUV for the least rugged sections of the road, which ran through a village and skirted the grounds of the Catholic diocese. St. Mary's, founded in 1959, had had the good fortune two years later to acquire the services of a unique medical couple, Italian doctor Piero Corti and Canadian physician and pediatric surgeon Lucille Teasdale. Corti arrived in Gulu first, then convinced Teasdale, whom he had met in Montreal, to come out for a couple of months to start a surgical department. They were married seven months later and would spend the rest of their lives running the hospital. Following the ouster of Idi Amin, St. Mary's was attacked and looted several times, but remained open for business, often treating the war-wounded. After a short hiatus, the hospital was once again targeted, this time by the LRA. In the midst of war, malaria and an epidemic of the hemorrhagic fever known as Ebola, which claimed the lives of twelve hospital staff, Teasdale performed more than thirteen thousand operations, during which she suffered the small cuts that doubtless led to her death in 1996 from HIV/AIDS.

As I waited for Victor in the compound, one of the religious fathers told me he'd been there for twenty-five years tending the equipment, a job that required his attention day and night.

"Why do you do it?' I asked.

He laughed at the question. "I was never good as a mechanic of souls. What else can I do?"

Victor was tall, lean and casually dressed, wearing an orange T-shirt with horizontal brown stripes and a wonderfully disarming grin. All my fears and hesitations about the visit to Gulu fell away.

"I knew you'd come." He shook my hand vigorously and ushered me down a long corridor, greeting patients, asking each about an injury or a relative, stopping to pat a child on the head. Everyone seemed to know him.

"We'll have a look at the ward first."

After I'd met his associates in the African Youth Initiative Network—Richard Onen, Peace Amuga and Jackson Opio—Victor led me into a small post-op ward with only six beds, so crowded that the floor was awash with women and children. A woman in a blue floral dress with a hem of white embroidery looked away shyly to avoid my gaze at her bandaged neck, where, Victor explained, a vicious knife wound had just been mended. Several children had at least one hand wrapped in gauze, a result of the tissue damage or worse they'd suffered when the LRA set fire to their houses. I was down on my knees fighting back the tears when one small boy, not yet five years old, broke from the group and put his arms around my neck.

He had lost both hands.

Victor was on his knees, too, hugging several children. "We've had a team of Dutch doctors here for two weeks doing plastic surgery. The work they did was fantastic." These were not all new wounds. Victor and his co-workers had scoured the towns and villages in a hundred-mile radius looking for people in need of medical attention. Out of the 185 individuals gathered in the first sweep, 95 had been selected for surgery; 72 of those surgeries had already been done. Some of the injured and their families had been so traumatized and discouraged it took several visits and the entreaties of friends to convince them the offers of help were genuine.

The compound consisted of two dormitories, an outdoor kitchen, a latrine and a washhouse in a large, treed expanse of five or six acres. We paused for lunch in an open concrete structure, where food was served from large tubs to patients and staff. We lined up, cafeteria-style, and the mixture of vegetables and tilapia, as well as *matoke*, a staple made from small green bananas that taste like potatoes, was heaped on our plates with millet bread in the shape of chapatis. Victor introduced me to everyone. Katherine, a woman in her eighties with closely cropped grey hair and a dress intricately patterned in bright yellow, blue and purple, was recovering from surgery on her neck. The LRA intended to hang her, but she had resisted so strenuously they slashed her throat instead.

"They missed my jugular," she rasped, grinning broadly. "Imagine, I was saved by a dull knife!" All she wanted now was to get home in time to climb her tree and pick the ripe mangoes.

Michael, a small boy who had been shot in the thigh four years earlier, stood listlessly, holding his father's hand. Although the bullet had been removed, his leg was still oozing pus. Continued infections had rendered his knee joint inoperative and threatened to make removal of the leg necessary. Michael was one of three hundred bullet and shrapnel cases Victor had identified as requiring immediate attention, and doctors had suggested an operation at Mulago Hospital in Kampala. Penniless, Michael's father, Simon Okot, was at his wits' end.

Before lunch, I'd been introduced to Evelyn, with whose misfortunes this chapter began. She had spoken to me under duress, as a favour to Victor, and did not stay long after relating her story. I looked for her in vain at lunch. I'd last seen her disappearing into one of the dormitories, where she would remain for the rest of the day. I left my food mostly untouched, thinking of what she had endured, and for which no simple remedy would be found. Victor made as if to ask what was wrong, but then thought better of it. Instead, he walked with me to a large tree in the middle of the compound, where he had cajoled a woman named Nancy into joining us.

Nancy took the plastic chair opposite me, Victor sitting off to one side. There was some awkward small talk, in which I offered a bit of my personal history—the writing, the preoccupation with human rights. She had been on the road when Kony's rebels attacked, Nancy said. They killed several of her companions, but because Nancy was pregnant they cut off her ears, nose and lips with a razor instead. "Give these to the government soldiers," they jeered, closing her fingers over the bloody pulp. Unlike Evelyn and the others, Nancy did not take her eyes off me while she spoke. Her fiercely defiant expression seemed to say: I dare you to look at this face. Over the next hour, I plied her with questions.

"*Waylaid* is the word Victor likes to use," she said. "I was waylaid on the road. Why? I never understood the logic. After they'd finished with me, I felt no pain; I was dead. You can imagine the rest." It was true. I could not stop imagining the things she left out. "My husband, a government soldier posted elsewhere, couldn't bear to look at me. He has another wife now. We Acholi have a saying: 'A poor mother is always better than a rich father.'"

To break the tension and to cover up my own distress, I mentioned a favourite line in Sean O'Casey's play *Juno and the Paycock* where, after the young Irish woman named Mary has been made pregnant and deserted by an English solicitor, she is comforted by her old mom with the words: "And what's better than a father? Two mothers."

Victor grinned and translated for Nancy. She nodded her head, then began to speak of the difficulties of making a living and looking after her daughter. Not just the problem of finding work, but dealing with the humiliation experienced by the young girl, who was teased and rejected by the other children. As Nancy talked, I could feel a tsunami of rage building in me against Joseph Kony, for all those murders, mutilations, lives destroyed. A bullet would be too good for him. He needed to suffer. Slowly. I wanted to know how someone like Nancy could deal with her anger or recover from what had happened.

Nancy thought for a moment, twisting the hem of her skirt in one hand. "I had to let it go, I was hurt enough already."

"What about justice?" I shouted, too troubled to keep my voice down. All the senseless deaths and violence I'd heard about here and in Rwanda had come flooding back. "What should happen to the people who injured you and killed your friends?"

"Justice?" Nancy extended her hands in a questioning gesture. "What's that?" No one spoke. In the silence that followed, she raised the fingers of one hand and gently touched her upper lip, which the Dutch surgeons had been trying to repair.

"Restorative surgery is what they call my operations. That's what I wish for Kony's people—restoration. Bring them back, integrate them into our community. They were mostly abducted boys."

I could not hide my shock. I was shaking from the interview, unable to control my breathing. The late afternoon sun had slipped behind the trees in the compound, sharpening the colours. Noticing my discomfort, Nancy graciously changed the subject. "Your first name sounds like the Acholi word for bicycle," she said. I was grateful for the reprieve. With Victor translating, I replied that being mistaken for a bicycle wasn't so bad, since my name in Japanese sounded like the word for diarrhea. They both laughed.

"What will you do now?" I asked her.

"I plant crops, mostly groundnuts, but what I want to do is go to school, study to be a nurse."

Uncertain how to end our conversation, I offered a foolish compliment on her dress, which was pink with a yellow collar. "Will you let me take your photograph?" Nancy looked at me for several long seconds before nodding her head. When her face appeared in the viewfinder, I closed both eyes for a second and released the shutter. I would return to that image often in the weeks that followed, pondering what it means to be human.

VICTOR HAD OTHER plans for the following day, so I spent a few hours exploring the town of Gulu, which was so much smaller than its turbulent place in the history of Uganda might suggest. I also needed to secure a return ticket to Kampala. Except for its bright colours, Gulu resembled a dusty frontier settlement with an unpaved main street bustling with activity and half a dozen intersecting dirt side roads. There were few trees. The only relief from the sun was to be found in the shade of buildings, which were, with the exception of several hotels, seldom more than a single storey. Motorcycles buzzed past on the hard-packed, rust-coloured earth, threading their way through the maze of pedestrians. A tall woman in an elegant dress and matching headscarf stopped to chat with friends outside Prince Electronics while her two children peered into the neighbouring pharmacy. A billboard from Crane Bank across the street advised opening a bank account NOW. Motorcycle repairs were conducted on the street alongside a barbershop with a corrugated tin roof advertising its services, and merchants grinding corn and filling huge sacks with the flour. Clothing, padlocks, goods of every sort lined the street on tables or hung from racks. Not far from a walled compound with a sign reading POLICE PRIMARY SCHOOL, three men were transforming scrap metal into pots, pans and braziers. Mama Cax's Beauty Salon had no customers, but several women sat sewing nearby amidst a heap of assorted materials. Between their stalls, a sign pointed to Alulululu Pork Roasting Joint & Cold Drink.

As I approached the main intersection, a woman in a pale blue singlet pedalled past on a bicycle, followed closely by a matching blue Isuzu truck carrying a dozen young soldiers, all eyes intent on the cyclist. The Eagle Air office was closed, but an obliging lawyer next door graciously went in search of the proprietor. Back at the Bomah Hotel, where I was staying in one of the small guest cabins set amongst palm trees and well-kept lawns—with an open circular central bar and restaurant—the real future of Gulu was very much in evidence. Less than fifty feet behind the bar with its attractive thatched roof loomed the spectre of the New Bomah Hotel, a six-storey concrete monstrosity surrounded by a spiderweb of spindly wooden scaffolding and totally lacking in character or elegance. Stalled for lack of funding, the new structure was meant to bring employment to Gulu and to provide digs for the steady procession of journalists, visitors and NGOs attracted to this troubled and hitherto neglected region.

The next morning, Richard Onen and I set out to visit Paicho IDP camp. Our taxi crept cautiously around ridges and gullies in the dirt road heading north from Gulu in the direction of the Sudan border. Christian music belted from the sound system. This was one of the roads taken by the famous night commuters, as many as forty thousand displaced children so frightened of spending a night in the inadequately protected IDP camps that they walked alone or in small groups every evening to the relative safety of Gulu, Kitgum or Pader, sleeping on streets and in vacant lots, returning home without breakfast or, if they were fortunate, with tea and a single biscuit. To highlight the plight of these kids, Canadian activists Adrian Bradbury and Kieran Hayward walked the seven miles into downtown Toronto every night for thirty-one days in July 2005 and slept in front of city hall. Since then, their organization's annual Gulu Walk has raised awareness and hundreds of thousands of dollars of support for these children.

As we drove, I recalled the argument advanced by Brazilian educator and philosopher Paolo Freire in *The Pedagogy of the Oppressed:* if you politicize language, give people the words to articulate their plight, name the sources of their victimization, they will learn in a few short months what it takes Western schoolchildren years to absorb. From what I'd seen so far, most of the good work on the ground, the effective aid, was being done by small groups of dedicated individuals like Victor and his friends here in Gulu and Peninah and her assistants at the Village of Hope outside Kigali.

With a modest grant from the victims' reparations program of the ICC, Victor and his AYINET colleagues had performed wonders. He had convinced Dutch surgeons to give freely of their time. His approach was grassroots and intimate. He knew the names and circumstances of all the people his group helped. I loved one story he'd recounted of his efforts to work with a demoralized group of people who had returned to their village to find their houses destroyed and the fields a mess. Drawing on his limited funds, Victor decided to offer each family forty pounds of high-grade rice, explaining that this special grain would be more productive than anything they had ever grown. "I know some of you will want to eat this rice," he said, "but to encourage you not to do that I'm going to give a free bicycle to the four families that produce the best yield." He returned shortly after harvest to find a dynamic and excited community, houses rebuilt, fields in good order, the yield generous. "I have a confession to make," he told the assembled group. "There are no bicycles. I lied. It was the only way I could think of to convince you to make the effort." Each family had produced much more rice than they needed to survive. "The secret," Victor said, "—and this is no lie—is not to sell your rice immediately. Have someone keep an eye on the market. Put away what you need to survive and what you need for the next crop. Only sell your surplus when the price is high. In this way, you'll soon be able to afford not only your own bicycle, but also a cow and a goat." The villagers took Victor's deception in good spirits, slapped him on the back and invited him to join them in a harvest feast.

It was on a road like this one that Alice Adong, too, had been waylaid. Like my conversations with Evelyn and Nancy, our interview had taken place the day before in the shade of a large eucalyptus in the compound, with Victor there to comfort and to translate. I could barely endure the pain in her eyes as Alice slowly shared her terrible story. "Rebels caught us in the garden between the cabbage rows and sorghum. The men, including my husband, were grabbed and beaten, the rest of us forced to crawl up a rocky incline until our knees were shredded to the bone. On pain of death, I killed my husband while the children watched." Her words struck me with a force so physical I had to brace myself on my plastic chair. A line from Shakespeare I had not thought of for twenty years emerged from the fog of memory: "As flies to wanton boys are we to the gods; They kill us for their sport." Aid workers had fled the village, Alice was saying. "A second group of rebels killed my mother and elder brother. The swelling was painless at first, then my leg grew large and hurt like hell. I got another man and two more kids, but then he left." Here she stopped. She seemed to be looking through me for a moment, into the hollowness at my core.

"My son dropped out of Senior One," she concluded. "Now he burns charcoal."

I was floundering in the web of violence she had described, trying to understand how anyone could withstand such suffering, such pain.

The question that burst out of me was addressed not so much to Alice as to the world in general: "What kind of people would commit such acts?" I knew the answer before it was spoken:

"People just like us."

In his essay "African Aesthetics—The Acholi Example," Okot p'Bitek explains the term *odoko dane*, which means, literally, to become human. The concept refers to young people who have learned the values of their culture, including generosity, good conduct and fine manners. In other words, they have grown up. Perhaps Joseph Kony, the scourge of Gulu, is someone who never grew up, I speculated. He takes refuge in the trappings of a foreign religion, calls upon foreign spirits for succour, but lacks the virtues associated with his own heritage. A man who could commit or condone such atrocities is not only *lapoya*—that is, insane—but also *kite pe*, without manners, a nothing, *pe dano*, less than human. I hung my head. With the stories of these women so fresh in my mind, I had been thinking only of punishment. As a mere observer, I was consumed with thoughts of retribution and revenge, whereas Nancy and Alice counselled forgiveness. I recalled the words of Bishop Desmond Tutu, a central figure in the Truth and Reconciliation Commission in South Africa: "Retributive justice is largely Western. The African understanding is far more restorative—not so much to punish as to redress and restore a balance that has been knocked askew."

According to Paicho headman Ajok Alphonse, the camp had been set up in 1996 in response to LRA attacks on surrounding villages, holding almost twenty thousand people at its peak. Rebels regularly attacked Paicho and other camps, looting, and killing and abducting children. Now that the Gulu region appeared more stable and the World Food Programme had cut food supplies to once a month, attempts were being made to send people back to their farms and villages. But dismantling the camps was easier said than done. Fear still prevailed, and the comforts, however minimal, of living in a camp with provisions and a modicum of security had made the generation born and raised there reluctant to go back to farming. Parents complained that discipline was an issue, kids running loose all day with no responsibilities, Ajok said. Alcoholism in the camp was a problem, with several brewing operations in full swing, using millet, cassava and sorghum to make beer. These kids had nothing to do in the camp, especially those child soldiers who had escaped from the LRA. No trauma counselling, no work, no literacy programs. The Ugandan government was doing next to nothing.

During our rounds of Paicho camp, we attracted clusters of children, curious about the gangly *mzungu* and eager to check out their images on the screen of my digital camera. Many were dressed in rags and hand-me-downs, without shoes. Others sprawled bored in front of the family hut or carted water on their heads in plastic jerry cans. Many young girls had infant siblings strapped to their backs. It wasn't until the tour was coming to an end that Richard stopped in front of a tiny mud hut with a thatched roof and padlocked blue metal door and announced that this was his home. His family had managed, against impossible odds, to get him an education, including basic medical training. In addition to his work with Victor and the others at AYINET, Richard also ran a tiny pharmacy no bigger than a closet on the main road through Paicho camp. This initiative earned him a few dollars and supplemented the camp's medical facility, which served five hundred people per month, three hundred of them repeats, suffering from malaria, respiratory problems, diarrhea, eye and ear infections and a constant shortage of drugs. When I asked the three beleaguered medical staff what they would like the outside world to know about their situation, one mentioned the need for buildings, equipment, a vehicle; another spoke of the high number of incidents of sexual abuse in the camps and the AIDS epidemic. I could see that their problems were systemic, the product of continued government neglect.

On our return journey to Gulu, I asked Ben, the taxi driver, to turn off the radio, which was once again blasting out Christian inspirational music. I wanted to talk, solicit his opinions. Why had the rebel attacks not been stopped, why was the north being neglected and its agricultural land not put to good use for the nation? What was needed to bring the Acholi back into the political process, to lead them towards economic recovery? When no answer was forthcoming from Ben, Richard interjected that people needed information about crop rotation, market variability, oxen, micro-credit. Despite scandals involving government ministers, high-level convictions seemed rare. Why was that? I persisted. As we jockeyed down the road, dodging ruts and potholes in the red earth, I expected an earful of the old animosities—the NRA's betrayal of Okello, the train full of civilians in Mukura set on fire by newly incorporated rogue elements in Museveni's conquering army—so I was surprised when Ben caught my eye in the rear-view mirror and offered this cryptic assessment: "Only if you kill will the law respond in Uganda. Otherwise, you can buy your way out of anything."

Museveni had remained unpopular here, receiving less than 25 percent of northern Uganda's support in the previous election. The Ugandan army had been both rough in its treatment of the local population and lacklustre in its pursuit of the LRA. The government was apparently content to let the Acholi be punished for their intransigence and meagre support, staging occasional peace talks it knew would come to naught.

Back at the AYINET compound, Katherine waved as she emerged from the dormitory. Her stitched throat was still mending, so the mangoes would have to wait. I bade a reluctant farewell to my friends, promising to keep in touch.

A dinner party to which I'd been invited was in progress at the Bomah Hotel. Aid workers from the Refugee Law Project had gathered to give Erin Baines, a law professor from Vancouver, a rollicking send-off. Julian Hopwood, an expat Brit, filled me in on plans for a proposed new sports complex for Gulu, with soccer coaches projected to visit from around the world. Joel Legatt, a Canadian filmmaker from London, was planning a documentary on the project. Several hotshot soccer players, they told me, were former child soldiers. Sport and culture: tools of reconciliation and healing. When I asked Erin what she thought about justice in Gulu, the woman next to Julian, who had been listening to our conversation, leaned across the table.

"You're asking about justice? Try this on for size. One of Kony's people recently indicted by the International Criminal Court is a former child soldier from this area, abducted at a tender young age. Now, overnight, to celebrate his eighteenth birthday, he is suddenly reclassified as a deadly adult war criminal. Where's the justice in that?"

five *The Moving Toyshop*

IT WAS the second day of action in the DRC's Rutshuru area. Militia units, having surrounded the town to ensure there was no armed resistance, began the house-to-house cleanup, seizing food and valuables, killing males of fighting age and taking their pick of women and girls. Kadima, who had days before been promoted from porter to fighter and equipped with knife and AK-47, trembled as the leader of his team of eight, a boy only two years older than him, pounded on the door of a ramshackle dwelling at the edge of town. As his blood, fuelled with drugs and adrenalin, drummed in his ears, he tried to focus. Failure could mean death or punishment. The leader struck the door again. The door opened a fraction and a frightened face appeared in the slit. The leader applied his shoulder to the door and burst into the room, thrusting his knife into the belly of the half-naked man, who crumpled to the floor. Kadima, pushed from behind, stumbled over the writhing form in the doorway, and found himself face to face with a young woman trying to cover herself with a sheet. She screamed and reached out to her dying husband, but the sheet was ripped away and she was tossed onto the bedclothes in the corner.

The leader, while shouting directions for removal, dropped his rifle and, unbuttoning his pants, moved towards the woman. She wept, but she did not resist. Then it was Kadima's turn. He gave the leader a thumbs-up, fumbled with his zipper and fell on the woman. She closed her eyes, then turned away. Kadima was limp, but imitated the movements of the leader, hoping his deception would not be noticed. The woman turned her head slightly and tried to see his face in the darkness. Before he could roll off, he was dragged aside by the next member of the unit. He reached out a hand to help himself sit up but recoiled in shock from the sticky fluid on the floor, oozing from the belly of the dead husband.

I lifted my fingers from the keyboard and looked out the window. A rusty hulk, belching a cloud of black smoke, had come into view in the outer harbour; it cut its engines and glided slowly, almost imperceptibly, towards the dock, several figures at the bow and stern preparing to secure lines. My own lines needed securing, too. There was no point trying to reconstruct a scene based on my interview with the female victim of these atrocities, or the myriad accounts I'd read by former child soldiers. Attempts to dramatize what I'd been told failed to capture either the violence perpetrated or the horror experienced, and served only to blur the line between fact and fiction. Yet the question troubled me as I wrote, because memory is unreliable and language is unruly, a medium that never says quite what you want it to say. Story, too, makes its own demands, pushing beyond fact for what has been called the deeper truth.

After three months back home reconnecting with my family and trying to make sense of my time in Rwanda and Uganda, I was back in Africa, in the Kivus to be exact, an Armageddon in the making, and it had taken all the energy I could muster to board the plane for my return. In short, I was afraid. Not just for my life, but that I was not up to the task of telling the stories with which I'd been entrusted. This had happened to me once before, while I was doing research for a book about prisoners of war in Hong Kong during World War II. After travelling to the crown colony, crawling through trenches, studying archival copies of the *South China Morning Post*, interviewing veterans who had survived the POW camps and reading everything I could find on the subject, I began to have nightmares, prisoner-of-war dreams. Fearing I was headed for a nervous breakdown, I abandoned the project for several years.

I could have flown directly to Goma from Entebbe, but flight departures were unreliable and a trail of lava from the Nyiragongo volcanic eruption seven years earlier had severely shortened the runway. It had never been repaired. One of the reports I downloaded from the Internet contained photographs of a cargo plane that had overshot the runway in Goma and exploded in a commercial district. A couple of the photos showed Congolese men trying to douse flames in the charred wreckage with water from plastic dishpans. Instead, I had caught a flight from Addis Ababa to Kigali, where I would make the short journey to the border by bus.

With an evening to spare, I decided to call one of my contacts, Philip Winter, and invited him to dine. We met at the Hotel Gorillas. He, too, was heading to Goma, where he worked as a consultant with the United Nations peacekeeping force. Africa-born and Cambridge-educated, Philip had managed a shipyard on the Nile at Juba, worked for Save the Children and Operation Lifeline in Sudan and served as executive assistant to the president of Botswana, a career trajectory that had kept him mostly in Africa, though he maintained a flat in London as a safety valve. When I asked what kept him here, he joked about having "salt balls" from trying to keep one foot in England and one in Africa. I said that resembled my own situation in Canada, having one foot in the Prairies and one on the Pacific coast, which left me buggered by the Rockies.

Philip was more forthcoming about his work for the UN. "Peacekeeping, peacemaking and peace enforcing, the whole gamut. I hate the rhetoric of the trade, but that's about it. The UN has no clear mandate, which makes it less effective than it should be. And no human rights component. All this stuff is 'partnered'—another of those ghastly nouns conscripted to do double duty as a verb—yes, partnered to NGOs."

He paused for a drink. "It's an ailing organization, but it's the best that we have."

Philip loathed the corruption in Kenya, his other part-time abode, and what he considered the absence of a sense of service in too many African leaders and bureaucrats. "This, not colonialism or foreign companies, is what is ruining things here."

"One of my favourite lines," I said, pausing to grind some pepper onto my grilled chicken, "is from the correspondence of W.B. Yeats. It's a quote from Montaigne, who says, 'A prince must sometimes commit a crime to save his people, but if he does so he must mourn all his life. I only hate the men who do not mourn.'"

"Yes, brilliant. They don't mourn." Philip topped up both wine-glasses. "Joseph Kabila, the DRC's president, may not be very bright, but at least he's aware of the fact and has the people around him that he needs."

I asked Philip why the joint Congo-Uganda operation against the Lord's Resistance Army had been such a disaster, dispersing the killers into small units that were taking revenge on the local population. After a Christmas Day massacre, the attackers had sat amongst the bodies of the dead and consumed the prepared feast.

"The idea was good: remove Kony and the LRA will disappear. The problem? Intelligence failure, bad weather and no Plan B." A country the size of Congo was impossible to police.

What about the so-called Dutch Disease, I wondered, the theory that resource windfalls often cause economies to collapse, agriculture being first to go, then manufacturing, so the country ends up poor and totally dependent on foreign aid and imports?

"This did not happen in Namibia or Botswana. Both are blessed with important natural resources, but have managed to function and foster civil society."

I wanted to ask more questions, but time was running out. The other patrons had departed and the waiter kept dropping by to ask if there was anything else.

"What's the solution for Congo?"

"Things work differently in Africa," Philip said, folding his napkin. "You can't expect the same attitudes to prevail here. Or depend on the usual solutions." To illustrate, Philip told me about an old man in Southern Sudan whom he found scratching his head as he looked at the sign outside the office of Save the Children. The old man wanted to know why so much emphasis was being placed on children. Philip gave him the standard Western explanation, that children are the hope for the future. The old man screwed up his weathered face and offered his own perspective on the situation. "We can always produce more children, but if we lose our language and traditions they're gone forever."

Philip encouraged me to phone him if I had trouble getting into DRC. A visa application, he warned, might take a couple of days. I paid the bill and accompanied him to the lobby, where the head and torso of a large plaster gorilla sat on the countertop. Philip shook my hand, then patted the animal's distended belly.

"You'd make good use of your time by going into the mountains and visiting this fellow."

Animal safaris were the last thing on my mind as I arrived at the border crossing at 9 AM in Gisenye. When I'd slipped into Congo from Rwanda a few months earlier, DRC had been at the centre of the world's attention. The rebel forces of Seventh-Day Adventist Tutsi general Laurent Nkunda had swept through the eastern province of North Kivu, overrunning villages, a government military base and IDP camps. Countless inhabitants were killed, women and girls gang-raped, shelters looted and burned, children carried off and trained to kill. Vast numbers of people had been driven onto the roads and into the jungle in search of refuge. Although Nkunda had been arrested and detained in Gisenye, and his forces integrated into the Congolese army, Goma was vulnerable to both the rebels and retreating government soldiers, most of whom were unpaid, poorly trained and dangerous. MONUC, the UN forces whose rules of engagement did not allow them to fire unless fired upon, had proven useless and become the target of insults and rocks hurled by infuriated residents.

The fixer I had hired was waiting for me at the border, dressed in a grey-and-white striped cloth cap, black jeans, a faded blue denim jacket and a black T-shirt on which I could detect faint outlines of the Charles Bridge in Prague. Célestin Kakule Kiza was the youngest of a Congolese family of eleven, his father an engineer who had once worked for Gécamines in the Katanga region. Célestin started learning English from an uncle who had worked in Zambia, then spent a year studying at an English centre. Having earned a degree in computer data processing, he tried to do further study in the field of management but dropped out because of a lack of money. Although he looked too young to know the ropes and to make the contacts I needed, I was relieved by his youth, which would allow me to play things by ear rather than feel embarrassed—as I would have in the hands of a more experienced fixer—about my ignorance and lack of a clear plan of action. We'd make an odd couple, but capable, I hoped, of winging it in the troubled Kivu provinces.

The first matter in need of "fixing" proved insurmountable. Despite Philip's warning, I had expected to pay the fee at the border and walk across as I'd done previously. This time, the customs officer demanded a pre-arranged letter of permission. No amount of cajoling registered on the stolid official, who had obviously seen too many *mzungus* of late and had better things to do than explain the new entry requirements. I directed my hangdog gaze at the woman seated in the corner who had processed my entry months earlier, but she could hardly be expected to remember me. I knew from Philip that two American journalists had been arrested in Walikale the week before for not having proper papers. The new regulations obviously had something to do with the current troop movements. Rwandan and Congolese armies had collaborated to bring Nkunda's forces under control and drive out the Hutu *génocidaires* of the rebel group the FDLR (Forces démocratiques de libération de Rwanda), now widely regarded as the enemy; and the Ugandans had joined Congolese forces to flush out the remnants of the Lord's Resistance Army from its bases in the north. It was rumoured that President Joseph Kabila had made a surprise visit to Goma to drive home the point that he was working to produce peace in the region and was well aware of the desire of certain parties, some of them African, to balkanize his country, which was not doing well in the housekeeping department, with countless dead and millions of dollars' worth of Congo's minerals being looted daily by foreigners and corrupt officials.

Célestin and I hung around the visa office for an hour, hoping for an official change of heart. Two French journalists were in the same predicament, talking a blue streak and gesticulating wildly. The stolid official departed for tea, but not before handing me a slip of paper with the e-mail links for a visa application. I bade a reluctant goodbye to Célestin and walked back into Rwanda to look for a hotel. However, after firing off the request form and requisite visa information, with a possible two days' wait in store, I decided instead to take Philip's advice to visit the gorillas.

I stayed overnight near the park offices in Kinigi, a small town in Rwanda's Musanze district, a bus ride of an hour and a half back along the main road from Gisenye, where I found a room in a small hotel that catered to gorilla trekkers. Early the next morning, I was assigned to a group of nine, including two Israelis, four Australians and a couple my own age from Texas. Our guide drove us two miles to the park entrance, where hiking sticks and porters were available for hire. Some of the porters had previously been poachers, I'd read in the brochure, but they had been persuaded to shift their allegiance as tourism increased and the number of gorillas dwindled. Although I had nothing but a camera, a sandwich and water to carry, I felt obliged to hire one of the porters, who handed me a stick with a beautifully carved handle and expressed disappointment when I insisted on carrying my own bag. After a twenty-minute briefing about the similarities between ape and human DNA and the dangers of transmitting colds and flus, the guide showed us a gorilla family album and gave advice about how to behave in their presence: no food, no eye contact, photos allowed but no flash, maintain a distance of seven yards from the animals.

I was ill-equipped, having left my gloves, raingear and hiking boots at home, but was determined to appear as spry as the youngest members of the troop. I set off ahead with the guide, hopping around mud holes, dodging protruding roots and branches and trying to conceal my laboured breathing as we climbed higher and higher. At the first rest stop, I made what I thought was an offhand joke about snakes. The guide laughed and announced to the others, "There's always someone in the group who asks about snakes. We try to guess who it will be." He patted me on the shoulder and added: "There are no snakes to worry about here." That's when it occurred to me I'd sunk five hundred dollars into an adventure that was not only going to embarrass me, but might also cost me a broken ankle or worse. I had visions of trekkers, guides and gorillas having a good laugh as I was hoisted by ropes into a medivac helicopter.

For the first hour, there was a lot of slipping and puffing, with minimal conversation. The guide made several phone calls to other staff stationed deeper in the jungle, who were tracking the gorillas and could advise us of their whereabouts. When we reached the flat area the gorillas usually inhabited at this time of day, they were not there. Another phone call, and we learned the animals had climbed farther up the slope where tender bamboo shoots were more plentiful. Leaving our bags with the porters, we made a slow, precarious ascent, traversing muddy slopes, sometimes losing our footing and having to grab a root, branch or extended hand.

Mountains and jungle stretched as far as the eye could see, palms, broad-leafed lobelia and what looked like giant marijuana plants. My heart was pumping overtime, responding to age, cramped muscles, a combination of fear and excitement and the mixed press gorillas had received. Stanley and early explorers had painted these creatures as deadly enemies to man, pounding their chests and letting out blood-curdling roars prior to attacking. Dian Fossey, buried in the vicinity, had painted quite a different picture, closer to the King Kong version, of gorillas as sensitive, peaceful creatures, violent only when threatened or defending their territory. As far as I was concerned, the jury was still out on the matter. As I was pondering the impossibility of a quick departure in such dense and rugged terrain, the guide raised an arm requesting silence and began to make a series of grunting noises to indicate we were friendly, just a group of not-so-distant relatives dropping by for a chat. He pointed toward a black and brownish mat half hidden in the dense foliage. The young female, clutching a slim green bamboo stalk in one hand, pulled herself into a sitting position, the other hand scratching behind her ear as if pondering the meaning of existence or, more likely, the presence of yet another band of intruders. By the time

I aimed the viewfinder, she had disappeared. To my left, moments later, a hairy arm extended from the greenery to clutch a branch. The youngster in question, partly concealed in the foliage, showed little interest in his guests, whether related or not, preferring to roll on his back, snatch the occasional leaf and rub his genitals. Sitting up with a half grin, he squeezed his broad, scooped nostrils between thumb and forefinger, supported his head in a primate's imitation of Rodin's *The Thinker*, then closed his eyes and played dead.

In contrast, nine trekkers, clicking cameras and whispering excitedly, jostled for position. A large teenaged ape lurched into view, knocked bushes aside and, taking advantage of a strong branch, muscled his way downhill. This young black-back male, the guide explained, would probably stay with the group until the older alpha male grew weak or died. The Texans, husband and wife, old hands at gorilla trekking, stepped aside to make it easier for the rest of us to take photographs. A mother and child I recognized from the family album sidled off into the bushes, stopping only when the mother heard the guide's familiar gruntings. The baby rolled back in her arms to look at us for a moment, its bright, brown eyes intensely curious, before retreating into the protective embrace.

Then Gohunda, appeared, all 550 pounds of him, an enormous black head and patch of silvered hair extending down his back from shoulders to hips. He knew we were there, but showed no fear as he lumbered towards me, parting the foliage and brushing branches aside. I held my breath, weighing the merits of various escape scenarios, yet wanting to record every detail. How to do this and avoid eye contact? As he advanced to within ten feet of me, I struggled to remain upright on the muddy incline, my camera stupidly poised. So much for the recommended security zone. From his erratic motions, I could not guess Gohunda's intentions, though they did not appear hostile. I tried a few grunting noises that sounded more like a walrus than a primate. Apparently unfazed by my accent and limited vocabulary, Gohunda returned my greeting. The only other sounds to be heard were wind in the trees and the occasional click of a camera. Then his strategy became obvious: he had positioned himself between me and a female and her baby. He pulled a branch through his mouth to strip it bare of leaves and gave me a thorough once-over: Old Silverback sizing up Old Silver Head. I felt honoured to have his attention. No, it was more than that, the genetic changes of three million years moving back and forth along that steady gaze, with so little accomplished at my end. I managed a few shots as he turned his back and moved off a few feet, where he hung onto a thick vine while processing the information, his free hand resting on a generous belly. He scratched his head, paused to let the obvious truth sink in—no threat here—then, with a final dismissive side glance, grasped an overhanging branch and swung himself downhill with surprising agility. The entire family disappeared into the bamboo below, where there was good shade for an afternoon nap. Following at a safe distance, we picked our way down to the grove, where we encountered a very pregnant female checking the wildlife in her armpit. Gohunda had already hit the sack.

Although I would not allow myself the luxury of sightseeing during this trip, events had conspired to make this encounter with African landscape and wildlife inevitable. The impact, after all the horrific human stories that were wreaking havoc in my brain cells, had touched me deeply, uplifting and saddening me at the same time. I rejoiced in what I had seen and learned, but was disheartened by the very real likelihood, given the resource wars and cross-border poaching, that this family of gorillas and their neighbours in DR Congo would soon be extinct. I had shared a precious hour with these gentle, secretive creatures, vegetarians all, moving with grace and innocence in their threatened habitat, only a few miles from where their human relatives were hacking each other to pieces. I bowed my head in shame and gratitude, with a special word of thanks to Philip Winter and the stringent new visa regulations in the DRC. Had any final illusions of being a more developed form of *Homo erectus* survived by the end of our trek, they were quickly dispelled on the way back to the parking lot when I caught my foot on a root and sprawled full-length in the mud.

I took the afternoon chicken bus back to Gisenye, not caring that I was the dirtiest foreigner in Rwanda. I showered at the hotel I'd found and spent an hour scraping mud from my shoes and scrubbing clothes in the sink. I had clean pants and one clean T-shirt left for dinner, but decided to have a swim first in the hotel pool. My body was in Gisenye, but my mind was still back with the gorillas and the families of farmers I'd observed in the surrounding area—men, women and children harvesting potatoes, their bicycles teetering along the road carrying huge plastic sacks of vegetables to market. A cyclist had leaned his bicycle-load of vegetables against a tree while he slipped into the bushes for a pee or a visit with his family. To keep the bike and its cargo from rolling down the modest incline in the road, he'd placed a single potato behind the rear wheel, a marriage of physics and faith.

When I found the visa permission letter amongst my e-mail messages, I was anxious to be on my way and called Célestin on my cellphone. I crossed the border without delay the next morning and dropped my bags off at the Bird Hotel, located in a dingy but quiet little enclave just off the centre of Goma, where he was waiting to receive me. We made the rounds of Goma on foot, changing U.S. dollars into Congolese francs, stopping briefly to set up appointments at UNICEF, HEAL Africa and the Regional Council of Development NGOs (CRONGD). The volcanic terrain was no less brutal than it had been at the end of November, but new construction was under way.

ALTHOUGH RAPE, ROBBERY and murder occurred with disturbing regularity, Goma's streets were considered safe in daytime. To the surprise of locals and the international community, the joint Congo-Rwanda military operation had resulted not only in the temporary dispersal of FDLR forces, but also in the flight and later capture of General Nkunda, still wearing his Rebels for Christ lapel pin and now reportedly under house arrest in Gisenye. The Rwandan government, embarrassed by a UN report that linked it with the financing of Nkunda and his renegade army, had decided to pull the plug on him. Its actions were no doubt influenced by the recent decision of Sweden and the Netherlands to suspend aid to Rwanda. In Nkunda's place,

Bosco Ntaganda had assumed command and overseen the merging of his CNDP (National Congress for the Defence of the People) combatants with the Congolese army. The arrangement was denounced by Human Rights Watch and various other organizations, especially the UN, because Ntaganda had been charged with war crimes and crimes against humanity by the International Criminal Court. For ordinary Congolese, the hiatus had proven short-lived; Hutu *génocidaires* and their followers were already drifting back to their former locations, taking revenge on the civilian population. No one knew yet how the situation would play out. Growing numbers of *génocidaires* had turned themselves in to the Security Council's DDRRR—disarmament, demobilization, repatriation, reintegration and resettlement—program for return to Rwanda. However, the number of returnees was meaningless if you considered the rate at which the FDLR was abducting children and forcing them into military service on a daily basis.

The staff at UNICEF seemed willing to consider my request to meet a few former child soldiers, whose readjustment they were facilitating in collaboration with MONUC and Save the Children. After a rigorous interview, in which I was grilled about my motives and aims, they agreed to introduce me the next morning to the coordinating team at the Centre de Transit et Orientation (CTO). I spent a couple of hours at the Doga, a hundred yards from the Bird Hotel, memorizing some basic greetings in Swahili. The waiter brought me an enormous bottle of beer. The bar advertised free wireless Internet, but the signal proved too weak to send or receive messages. The other patrons explained that the phone system in the eastern provinces was overloaded at the best of times. The beer, at least, performed as expected, and in due course my frustration gave way to laughter.

When Célestin fetched me in the morning, I was tired and irritated, not a good frame of mind for conducting interviews. I'd left my mosquito net by mistake at the hotel in Gisenye, and as a result, I'd spent a restless night in bird-land dreaming of violent encounters and waging a losing battle with mosquitoes in a room with ragged screens. I'd risen early to shift my belongings to a hotel down the road with the preposterous title of the VIP. As there had been no birds at the Bird Hotel, it seemed appropriate to find no very important persons staying at the VIP. The other guests all looked as dishevelled and down-at-the-heels as I did.

The training and orientation centre, temporary home to 150 former child soldiers, resembled a prison with huge walls constructed of jagged volcanic rock. The compound, an area of two acres, had no grass, only packed earth with a single forlorn tree in the middle, warped and bent from abuse and neglect. Clusters of bored and angry-looking young men hung out in corners or lay sprawled in the shade. Célestin and I were given a lecture by a senior staff member about the program: early wake-up each day for the residents, breakfast, dishes, Swahili class, science, math, free time, sports, games, plays, dinner, more dishes, skill training, basket weaving, painting, films, theatre, dance, music. It sounded dynamic and uplifting—a combination of community centre, health spa and arts club—but nothing of the sort was in evidence as we made the obligatory circuit of the compound, where food was being stirred in enormous pots over an open fire.

The first former child soldier ushered in to talk with me was a young girl, perhaps sixteen, I'll call Marie, who had joined a militia voluntarily when she was only ten years old. She thought it would be an adventure, she said. Marie was one of a family of nine girls. Her father was a pastor and she had sung in the choir. She joined the Mai-Mai, which later collaborated with other militias and Hutu rebels under the banner of PARECO (Coalition of the Congolese Patriotic Resistance), ostensibly in opposition to the depredations and threats of Laurent Nkunda and his forces. PARECO was involved in kidnapping, cattle raids and bloody attacks against its adversaries. First, she had served as a drudge, Marie said, carrying loads and washing clothes, with no shortage of kicks and prods thrown into the bargain. She complained that there had been no training available. When I asked if she had learned how to shoot a gun, she made a chilling remark: "No, we were not trained with guns; we used knives."

Had she been raped? It's likely, I thought, but I could not bring myself to ask such an intimate question of a complete stranger; it was the kind of experience you might not share even with a relative or close friend, especially in a milieu where victims of sexual abuse were often treated as pariahs and shunned by family and community. Instead, I inquired if there was a single experience that had affected her profoundly.

She hesitated, looked out the window. I followed her gaze across the barren compound, stopping at the steep wall of jagged volcanic rock, which evoked prison rather than protection. She shifted her weight in the chair and spoke in a barely audible voice.

"Seeing my best friend killed in battle."

"Was that what prompted you to escape?"

"No, I was homesick," she sighed, and looked down at her hands. "And I was afraid for my soul. I didn't want to go to hell."

The small room we'd been assigned for our interviews had open windows in which curious faces gathered every few minutes and had to be shooed away. The door opened and closed constantly as staff and other children shuffled from room to room. Célestin had offered to interpret, but wisely deferred to the female staffer who spoke excellent English.

The second former child soldier introduced to me, another girl, had also joined a militia voluntarily, at age eleven. I tried out a few words of Swahili. She remained silent, refusing to look at me. I tried introducing some experiences from my own childhood that had been violent and traumatic, including being stabbed in the stomach by a school bully, but my strategy did not work. She slouched in her chair, jaw set, arms folded across her chest, body language that indicated she was there under duress and wished I would piss off.

Things were not going as I'd hoped. I looked at Célestin, my raised eyebrows a request for help. He shrugged.

The interview ground to a halt. When I thanked her for coming, she suddenly found her voice, shouting. "I joined the militia because I envied people who knew how to use guns." She pushed back her chair, stood up and, placing her hands flat on the table, leaned towards me.

"My family doesn't give a damn what I do. Neither do you. And I'm going to rejoin as soon as I get out of this hole."

David, the third child soldier, had been abducted at age ten, though he hailed from the Masisi area, where there were plenty of reasons to have volunteered, such as poverty, danger, disease. He spent several years in the Mai-Mai, one of the bloodiest militias. He ran away six times, he told me through the interpreter, but kept being caught. He had numerous complaints about ill treatment—bad and insufficient food, lack of pay and physical abuse—but not a word about danger or having to kill people. He reminded me of Ishmael Beah and his memoir, *A Long Way Gone.* As a young teenager, Beah had become a highly efficient killing machine in his home country of Sierra Leone, all the more lethal because he was short in stature and could ambush the enemy from behind small shrubs. I thought of Nancy and Alice in Uganda, one mutilated by child soldiers, the other forced by them to murder her husband. At age sixteen, in the face of public hostility and an uncertain future, David was determined to hold his cards close to his chest. As with the girls, the psychological wounds he refused to discuss were evident in his face. When Célestin and I left the CTO compound later, I noticed David leaning against the wall by himself, staring into the distance.

As I sat with Célestin at the Doga, contemplating the bubbles rising in my glass of beer, I realized my expectations had been unrealistic. These kids, traumatized, abused and shunned, had scarcely opened up to the staff at the CTO after weeks of work to gain their trust. How naive I'd been.

"There's something strange in the air in Goma," I told Célestin.

He finished his chicken leg and placed the bone on his plate. "There is so much poverty and neglect here," he said, "if you aim for something better, or higher, you are suspected, envied and might even be killed for your efforts. It's not an easy place to get ahead."

Célestin's dream was to do postgraduate work in journalism and creative writing, so I spent the next half-hour giving him a few tips. I described how to put together a portfolio and explained the difference between talking *about* pain in a text and finding the words to *evoke* pain in a reader. Then it occurred to me that my companion had witnessed a lot more pain and violence in his few short years than I had seen in a lifetime, and that the Greek theatrical mode, referring to violence obliquely, keeping it offstage, might be the best solution for him. At that moment, the Doga began to shudder and we heard a huge roar a few hundred feet overhead. Oh, my God, I thought, the volcano. But it was only a UN cargo plane lifting off, its flight path directly over Goma's main drag.

In explaining why education was important to him, Célestin had used the quaint word *pedigree*. When I suggested that in English, *pedigree* is mostly used in connection with the upper classes or purebred animals, he waved the distinction aside. He wanted me to understand that family is important in Congo, that parents often have to choose from amongst many children the one who will be educated, the one most likely to live up to expectations, to bring honour and support to the rest of the family. He was the chosen one, the fifth born, so it was crucial that he succeed.

"I appear to have a bright future," he told me. "That's why my responsibility is huge."

He was reluctant to talk about his experiences as a fixer. He did, however, mention being on assignment in Kiwanja after a massacre, bodies strewn everywhere. The experience had made him sick to his stomach. The next day he overheard two young journalists obviously referring to him.

"Let's speak to that guy over there who knows where the bodies are. It's gonna be great, plenty to shoot!"

"Right on!"

SINCE NKUNDA'S OFFENSIVE, which had stopped at the northern perimeter of Goma, the number of displaced persons at the nearby Kabati camp had swelled to sixty thousand as terrified civilians fled their farms and villages. I was keen to go farther afield in the conflict zone and had previously contacted the Toronto offices of Médecins

Sans Frontières, requesting permission to visit its medical unit at Masisi in the hills west of Goma, an area where hundreds of thousands of people had been displaced by ongoing violence. The team was overworked and combating impossible odds. Although the hospital had increased its capacity from 72 to 175 beds, the staff were hard pressed to treat the distressed and beleaguered population and the wounded from all sides of the conflict. To add to these problems, cholera had broken out, and a landslide had killed fourteen people in the Kilimani IDP camp. Lack of food and adequate health care, in addition to the loss of crops and property, was destroying the morale of survivors, weakening immune systems and making treatment even more difficult. I had great admiration for MSF as an organization and wanted to see this team in action. I was curious to know what attracts doctors, nurses and logistical support workers to conditions where only the strict maintenance of impartiality keeps violence at bay. In one of MSF's medical blogs, a young doctor and psychiatrist from Toronto named James Maskalyk, stationed with MSF in Sudan, observed that a sense of "weightlessness" was one of the gifts of working under these extreme conditions. He meant the loss of ego, release from the self and all its baggage, that comes from working with a team of people committed to saving lives, often at the risk of their own. Add to that the pettiness of consumer society claiming your attention, and you have ample reason for choosing to live on the razor's edge.

Because of increased rebel activity in the DRC, I had been refused permission to visit the project in Masisi, but I managed a telephone conversation before I left Canada with MSF logistician Jake Wadland, who'd worked on the medical front lines in Africa. Jake was candid and down-to-earth. He recognized immediately the weightlessness syndrome.

"It's different for every individual," he said. "You get used to levels of stress, often to your own disadvantage, or that of the team. It's like a frog squatting in the water as it heats up, eventually finding itself cooked."

The challenges and disappointments of the job were legion. Government bureaucracy had botched the handover of one MSF operation in the neighbouring Republic of Congo; the hospital there had been looted by a single individual, destroying the painstaking work done to gain the trust of locals. In one of his blog posts, Jake wrote about seeing raw meat in the Talangai night market in Brazzaville, where you could buy turtle, weird-looking fish, snake and bloody slabs of gazelle with "the sales morbidly supervised by the gazelle's own recently severed head." Monkey meat was verboten whenever there was an outbreak of Ebola, Wadland's driver Euloge had informed him. Jake cranked up my hypochondria several notches by adding that the disease, "in extreme cases, causes your internal organs to liquefy and leak out of you."

Did he have any advice for me? "Camouflage your intent," Jake replied. Avoid any mention of justice and human rights. Instead, stand in the airport queue quietly and let the Israeli, Chinese or South African businessman you've just met describe how he conducts, with absolute impunity, his illegal logging or mining operations.

The day following my interviews at the CTO, I visited the headquarters of Save the Children France (SCF), an imposing walled compound in Goma with security measures that seemed more appropriate for the U.S. Mint. The only item out of keeping with the tight security was an old dog that slouched from spot to spot, looking for the most comfortable place to rest her arthritic bones. I could sympathize. While I waited for Célestin to arrive, I approached the dog, where she had settled on the veranda, extending my hand in greeting. Without warning, she bared her discoloured teeth and clamped them on my outstretched fingers.

I yelped, jumping backwards. The bite was modest—a warning only, the skin not broken—but it had the desired effect. I backed off. Célestin had arrived just in time to witness the attack. "I hope the children get a better welcome than this," he joked.

The two of us were greeted by an articulate young woman named Emmanuelle. She was from Lyon and worked as a child protection officer. Save the Children France received funding from UNICEF, the

World Bank and a variety of other donors. "But there's never enough money for the work that needs doing," she explained. Emmanuelle had endured tough working conditions in Yemen, Chad and Darfur, but DRC was definitely the biggest challenge.

"Living here is okay, but hope is in short supply. What troubles me most is the erosion of morality and the breakdown of civil society." Emmanuelle's earnest face assumed a pained expression. "There is a plethora of new actors on the scene, NGOs whose services are desperately needed throughout the region, but they refuse to locate anywhere except Goma."

I could appreciate her dismay, but Goma was the only spot in North Kivu that offered a scrap of security, that was not completely lawless. I made this point.

"Yes, yes, it's not just that. These competing agencies didn't consult or interact; each reinvents the wheel. MONUC agreed to collaborate with us on the matter of child soldiers, but only reluctantly."

I wanted to know what she thought about the perception back home that aid and intervention often seem counterproductive. Emmanuelle admitted that aid organizations make mistakes, but she rejected the cynicism of workers and observers whom she had heard saying of the local situation: "Let them kill each other." Her ambition was to work in aid and relief advocacy.

"Most UN advisers have spent only a few hours in the field. They have no idea about conditions on the ground or how these might be addressed. And journalists"—she rolled her eyes—"they write only about what is fashionable, and never in depth. I had a journalist in Darfur say to me: 'I want to interview a fifteen-year-old rape victim who's fluent in English.' It's voyeurism, pure and simple."

Emmanuelle paused, then added, "Well, maybe not so pure."

I mentioned the difficulty I'd experienced making a connection with the three former child soldiers at the CTO. Emmanuelle admitted to having the same problem. "Sometimes it takes months to earn the necessary trust, even for me." Foster homes are a better environment for helping these children relax and open up, she suggested. "The

CTO was meant for 50 children, but now has 150." She explained that there'd been a riot in the centre just that morning, from overcrowding and insufficient staff, and that it was under lockdown. As an alternative, SCF was sponsoring thirty-five foster families who received three dollars a day for providing shelter, food and emotional support for a former child soldier.

As we talked on the veranda of SCF, a journalist from the U.K. and his fixer arrived, wanting help to arrange a visit to the CTO. The journalist was harried, insistent; he had only two hours to spare. Emmanuelle explained about the riot and gave him names and a phone number, but did not mention the foster homes or offer the use of an SCF vehicle.

The following day, Célestin accompanied me to the faith-based HEAL Africa compound, just across the road from SCF. The huge complex, which began in 1998 as a modest, locally staffed hospital, was founded by Dr. "Jo" Kasereka Lusi, a Congolese orthopedic surgeon, and his English wife, Lyn, a social activist and administrator. Lyn had no doubts about the cause of Congo's problems; she knew it was not tribal conflict, but a resource war that fuelled the unrest and bloodshed. As she says on the HEAL Africa website, "This is the scandal of Congo: such a rich country, attracting predators from all the nations of the world, but with the majority of the people living in grinding poverty." She and Jo were determined to create a facility whose roots extended deep into the community, hiring and training Congolese from every religion and ethnic background. When the hospital was destroyed by the volcanic eruption of 2002, they rebuilt it from scratch.

Virginie Mumbere was a human dynamo whose work at HEAL Africa seemed to combine financial management, personnel, troubleshooting, public relations and counselling. Barely able to complete a sentence in her small office without a tap on the door or a call on her cellphone, she still took the time to show us through the grounds and the various wards and clinics, recalling the names of every patient and staff member. This helpful excursion also provided us some uninterrupted time with her. Much of HEAL's work, she explained, is doing fistula surgery. Women who have experienced multiple rapes, or been penetrated with sticks, broken bottles and gun barrels, often suffer damage to the lining separating the vagina and bladder, which causes the continual leakage of urine. In addition to the nagging discomfort and humiliation, many of these victims suffer rejection by husbands and family members. But the hospital's approach, Virginie assured me, offered a wide range of care and intervention.

"We don't discriminate," she explained. "And the treatment here is holistic. You can't just repair the physical damage; you have to try to heal the emotional damage as well."

She ushered us into a room where a dozen women were learning to sew on old-fashioned Singer sewing machines, the kind you'd find only in antique shops or second-hand stores back home. In addition to the healing experience of working together to create something of use and beauty, these women took away new skills to help them earn a living. Shirts, purses, handbags, computer sleeves and dresses were on display. Others in the compound were learning how to garden or to craft bowls, baskets and mats out of banana leaves. Staff and counsellors also made use of music, dance and film. HEAL, an acronym for Health, Education, Action and Leadership, clearly lived up to its name.

I'd asked Virginie if I could interview someone who'd recently arrived at HEAL, and she was on the lookout during our rounds for a patient she had in mind. A thirty-two-year-old woman I'll call Mbeda followed us reluctantly back to Virginie's office, where she slumped onto the couch, arms wrapped around her in a self-protective embrace, and concentrated on a badly worn spot on the carpet. Virginie, who spoke excellent English, agreed to translate. I was glad to be able to address my questions to a woman, especially one who radiated sympathy and trustworthiness. Célestin was glad, too; relieved, he melted into the corner for the duration of the interview. I began by saying that I felt very uncomfortable, as a stranger and as a man, to be asking for Mbeda's confidence, but explained that I thought outsiders needed to know what was happening to women in Congo.

"There was a knock on the door," Mbeda said slowly, each word hauled reluctantly into the light. "It was nighttime. 'Who are you?' my husband demanded. When he opened the door, eight soldiers burst into the room." She hesitated, as if the effort of recall were more than she could manage. "They killed my husband. All but one of them raped me."

I asked if she recognized the soldiers or knew what faction they represented, but she was not finished recounting her story.

"I lost my whole family. The rest fled into the forest and my parents died there. I don't know what happened to my brothers. I had to take care of their six children, but I had no home, no work, no food. How could I keep them from stealing to stay alive? I tried to have strength. One of my nieces went away to be a soldier." I thought of the two girls I'd interviewed the previous day and wondered what had really driven them into the treacherous arms of the militias.

"My house was destroyed. A woman in the village had a room she let us use and a sewing machine. Seven of us in one room. I could not imagine this kind of thing happening to me. Then it happened again. Two more soldiers raped me. And—" she paused and, with brows lifted, looked deep into my eyes, "my eight-month-old niece."

I did not know where to put my face. My mind went blank. After a minute, I blurted out a question about how she was recovering, what she was learning here at the hospital.

"I've had six fistula operations," she sighed. "My niece is ruined. I haven't menstruated since. They were Rwandan soldiers. CNDP, FDLR, what does it matter? I suffer high blood pressure, headaches. I'm afraid all the time."

And what difference would it make to Mbeda, I thought, if Thomas Lubanga Dyilo, Jean-Pierre Bemba or Bosco Ntaganda were convicted of war crimes and spent the rest of his life in detention in The Hague, eating three good meals a day, sleeping in a warm bed, watching a twenty-seven-inch television screen and playing Ping-Pong with Radovan Karadzic? Moments earlier, her hands, released at last from their tight grip on her shoulders, had started gesticulating wildly. Now, every blockage seemed to have been swept away and she wept openly. All of us were weeping.

I reached out to touch her. "What will you do? Where will you go?"

"I have no idea where the children are," she sobbed. She thought she should go back to Rutshuru, but it was dangerous in the bush. She had to be strong, she said, to help the children overcome their bad habits. If she could find a way to sew, she might earn a bit of money. The cluttered confessional space of Virginie's office fell silent, each of us exhausted and lost in our own thoughts. Miraculously, no one had disturbed us, though the hall outside was full of people waiting to talk to Virginie.

I asked how much a sewing machine would cost. "One hundred and fifty dollars," Virginie said, "plus maybe fifty dollars for materials."

When the interview was over, I made arrangements with Virginie for a sewing machine to be made available for Mbeda when she left the hospital.

Back out on the street, Goma had assumed an air of nonchalance, going on with what resembled business as usual. A fleet of small white clouds was drifting north, forming a path of celestial paving stones all the way from Lake Kivu to the smoking lip of the Nyiragongo volcano. Célestin, who had said nothing for the previous hour and a half, shook his head and declared, to me and to the passing clouds: "The abnormal has become normal here."

IT WAS NOT the Congo River, but maybe Lake Kivu would satisfy some of my watery aspirations. Bukavu, the capital of South Kivu province, would be my last stop in the DRC. Célestin, whose services I no longer needed, dropped a boulder-sized hint that he had never been across the lake and had a friend in Bukavu he hoped someday to visit, so I agreed to pay his fare one way on the fast ferry. We set off early the next morning for the harbour on the backs of two *boda-bodas*, only to be stopped by several Congolese soldiers who wanted a contribution to their beer and cigarette fund. Knowing time was short and they had probably not been paid that month, I gave them a few dollars' worth of francs and we all waved and laughed as the *boda-boda* engines were fired up again. Although we had reservations for the fully booked fast ferry, the French bureaucratic mindset was so firmly entrenched in Congo that it took more than an hour for passengers to fill out the half-dozen documents required for the two-and-a-half-hour passage.

As we pulled out of the harbour, a broad wake fanning out behind us and islands coming up in the distance, I could feel some of the weight of Goma lift off, until I met Jim Farrell, an itinerant Edmonton writer and journalist with a shaved head and a nose for danger. Jim had worked for the World Food Programme in Africa, he told me, and had ridden a Honda 600CC motorcycle across Europe to Afghanistan. Wasting no time on small talk, he launched into a description of "lake turnover."

"It's a loaded gun," Jim announced, with obvious delight. "Once there's an upward movement established, perhaps as a result of volcanic activity, the whole thing explodes like a shaken bottle of Coca-Cola."

He was referring to the tendency of rift lakes such as Kivu to blow up, the layers of methane gas and carbon dioxide at the bottom expanding upwards to release into the atmosphere, causing instant death to surrounding communities and all other living organisms. According to Jim's research, a massive extermination was likely every thousand years. Yet imminent threat seemed the last thing on the minds of passengers and crew gathered on the stern deck for a smoke, a bit of sun and some lively chat. Small dugout canoes drifted in and out of camera range, and the occasional steel clunker huffed past, barges in tow, its ancient engine churning out a spume of diesel smoke. Bogart and Hepburn in the cockpit of the *African Queen* would not have been out of place.

While I pored over my Goma notes in the main cabin, Jim Farrell and Célestin, both news junkies, indulged in journalistic shoptalk on deck. The ferry was now skirting beautiful islands, along the shores of which were moored a dozen small flat-bottomed craft, each with long, wispy spars, looking as delicate as an insect's antennae, extending from bow and stern for attaching a net. A woman in a bright red dress and matching turban set out from shore in a small dugout canoe.

Beyond her, several families were going about their business around eight tiny A-frame grass huts. Tranquility and neighbourly bustle prevailed out here on the water. The contrast made me think again of Conrad. After thirty-five years, my head was still full of scenes from *Heart of Darkness:* a European warship senselessly shelling the coast, slaves languishing in chains, the overturned railway car with its wheels upturned in a mute appeal to the heavens, human heads on stakes.

Conrad was one of my silent companions in the Congo. Several lines of poetry, too, kept coming back to me from Alexander Pope's eighteenth-century poem *The Rape of the Lock*, in which he satirizes the moral confusion evident in the mix of items on the fair Belinda's dressing table: "puffs, powders, patches, bibles, billet-doux." Using a word with the power of "rape" to describe the comparatively innocuous theft of a lock of hair was just one of the many ways in which Pope satirized his society. The perversion of spiritual values and the resulting emotional turmoil had stirred up what he called "the moving toyshop of [the] heart," resulting in serious moral breaches and other travesties of behaviour.

Rape was all too real in this huge country, almost the size of Western Europe, first known to outsiders as Belgian Congo, then Zaire, and now by the grossly inappropriate title of the Democratic Republic of Congo. Few people on earth had endured greater hardships at the hands of their own leaders or witnessed a more obscene and rapacious desecration by invaders. In the late 1800s, when King Leopold of Belgium decided to enhance his personal fortunes by pretending to offer the gift of "civilization" to this vast territory, he initiated a reign of terror, a deadly scramble for loot that continues to this day. To satisfy a growing world demand for rubber, Leopold and his gang of ruthless sadists used intimidation and violence to extract the maximum from his slaves. Workers who did not meet the required quotas had a hand or an arm lopped off. They were the lucky ones. Others were strung up, crucified, mutilated and left to endure slow, painful deaths. Millions were murdered or died of disease and starvation. The public outcry, including that by writers such as Conrad, Roger Casement, Mark Twain and R.B. Cunningham Graham, exposed Leopold's sham to the world and led to a temporary hiatus.

So much for figurative uses of the word *rape*. Still bathed in blood, as the battle for control of minerals rages, Congo has also gained a reputation as the worst country in the world to be a woman, a place where countless thousands have been sexually abused and mutilated. In many African countries, as elsewhere throughout history, the female body has become an extension of the battlefield. I'd lain awake all night at the VIP Hotel after my visit to HEAL Africa, the details of Mbeda's story an unending kaleidoscope of images in my brain. Her testimony had opened a floodgate in me, forcing me to confront my own relationships with women. I had been raised in a strict Baptist church and trained to think that sex, especially the premarital kind, was a landmine, to be avoided at all costs. The guilt associated with my urges—all the clumsy, adolescent fumblings—was a poor foundation for marriage. I was too embarrassed to discuss my ignorance and ineptitude openly; instead, I sought intimacy in affairs and one-night stands. This pattern continued through a second marriage, inflicting pain and humiliation. More than once, I'd washed and shaved in the dark so as not to have to look at my own face in the mirror.

I knew that sexual violence in Rwanda, Uganda, the DRC and elsewhere was not really about sex, but a tactic of domination and degradation—in other words, a weapon of war. A United Nations Security Council resolution, as recently as 2008, had set this interpretation down clearly. However, as I tossed and turned in my sweat-soaked bed at the VIP Hotel, I shrank from the memory of my own callous actions, which seemed less reprehensible only by degree.

So far, I had found plenty of evidence that the populace of Congo was deeply troubled and desirous of significant change. The day before we left Goma, I'd met with Kubuya Muhangi, who ran the NGO coordinating organization CRONGD and was active in various advocacy groups. When I asked him about the human rights situation in the DRC, he told me he had spoken with an old peasant who considered the current regime worse than Mobutu's; at least with the latter, you knew where you stood. Obliged to tell this story too often, Kubuya was resistant to my questions at first but soon gave a passionate indictment of the status quo in Congo: impassable roads, underpaid soldiers and civil servants, the law neither respected nor enforced, corrupt prosecutors, generals who built huge mansions in town and on the waterfront—where were the human rights in that? The government provided nothing for services; without foreign aid, he said, people would be helpless. He shuffled through some papers on his desk and showed me a map of eastern DRC, a huge chunk of which was still controlled by the FDLR, the Hutu *génocidaires.* Soldiers went unpaid as officials and politicians investing in real estate feathered their nests.

Kubuya's strong opinions kept him in constant danger. He was not content to give me the usual statistics on rape, murder and displacement; instead his focus was the root causes of violence. When I asked him about foreign interference in the DRC, particularly that of mining companies, he took a moment to reply, then shook his head: "We live and die in rags, but are buried in the richest soil in Africa." He bowed his head, then looked at me and said: "There's no shame, no shame."

Passionate, outspoken, honest, Kubuya was cut from the same cloth as Patrice Lumumba, the hero of Congolese independence. During his inaugural address, on June 30, 1960, Lumumba had set aside his prepared notes and openly denounced the Belgians, an honest and long overdue speech that infuriated the departing rulers who still had designs on the new country's resources. "Nous ne sommes plus vos singes," Lumumba declared to the assembled dignitaries. We are no longer your monkeys. The rest of the text is known word-for-word by most Congolese:

> We have known sarcasm and insults, endured blows morning, noon and night, because we were "niggers" . . . We have seen our lands despoiled under the terms of what was supposedly the law of the land but which only recognized the right of the strongest. We have seen that the law was quite different for a white than for

a black: accommodating for the former, cruel and inhuman for the latter. We have seen the terrible suffering of those banished to remote regions because of their political opinions or religious beliefs; exiled within their own country, their fate was truly worse than death itself. . . And finally, who can forget the volleys of gunfire in which so many of our brothers perished, the cells where the authorities threw those who would not submit to a rule where justice meant oppression and exploitation.

Although Lumumba was assassinated shortly afterwards and replaced by a ruthless dictator supported by Belgium and the United States, his legacy lives on in people like Kubuya Muhangi, Virginie Mumbere, Jo and Lyn Lusi and so many others.

Greed, racism and cruelty, the unholy trinity, had split the DRC asunder. The rape pandemic and ongoing slaughter were not unique to Congo; what was unique was the total impunity with which these things were happening. It would be convenient to call this the legacy of colonialism, to lay the blame solely on the outsiders, whose veneer of civilized behaviour was stripped away by the tropics. A passage in Louis-Ferdinand Céline's novel *Journey to the End of the Night* makes precisely this point. Even the ship carrying the protagonist south is rotten, held together only by numerous layers of paint. The passengers begin to unravel morally and emotionally the closer they get to Africa:

It didn't take long. In that despondent changeless heat the entire human content of the ship congealed into massive drunkenness. People moved flabbily about like squid in a tank of tepid smelly water. From that moment on we saw, rising to the surface, the terrifying nature of white men, exasperated, freed from constraint, absolutely unbuttoned, their true nature, same as in the war. That tropical steam bath called forth instincts as August breeds toads and snakes on the fissured walls of prisons. In the European cold, under grey, puritanical northern skies, we seldom get to see our

> brothers' festering cruelty except in times of carnage, but when roused by the foul fevers of the tropics, their rottenness rises to the surface. That's when the frantic unbuttoning sets in, when filth triumphs and covers us entirely. It's a biological confession. Once work and cold weather cease to constrain us, once they relax their grip, the white man shows you the same spectacle as a beautiful beach when the tide goes out: the truth, fetid pools, crabs, carrion, and turds.

The passage is overwritten with obvious zest, and in a mock-Conradian style. However, in its mockery, Céline's writing makes light of what is still a deadly business in Africa. Neo-colonialists peddle their lethal, high-tech wares and sow chaos under the umbrella of international agreements, with funding from multinational corporations and the blessing of local governments. Where once white traders and functionaries brutalized their African subjects, now the vested interests have turned this task over to the locals.

Jim and Célestin were still deep in conversation on deck as I put my notebook away. Sunlight reflected on the calm surface of the lake was dazzling. I watched the cluster of verdant islands drift astern, thankful for the soothing effect of the lake crossing. Water—even the deceptively placid surface of Lake Kivu—was always a spiritual medium for me, a springboard to meditation. I was feeling like the young MSF psychiatrist, emptied out, a state both liberating and deadly. André Gide, another of my literary companions on the trip, described this feeling in his memoir *Voyage to the Congo* as "l'oubli de soi total," a complete forgetting of self. In that liberated state, Gide felt he was capable of becoming all men, of entering into their very beings. However, he also realized that, in this emptied-out state, he contained the best and the worst of humankind, that the conditions necessary for the creation of art, and perhaps for the apprehension of truth, had their downside. "Il n'y a pas de différence essential entre l'honnête homme et le gredin—et que l'honnête homme puisse devenir un gredin, voilà le terrible et le vrai," he wrote. There is no essential difference between the honest man and the knave—and the honest man is capable of becoming a knave, that is the horror and the truth.

Welcome aboard, Mr. Kurtz.

All of the Congolese I'd spoken to agreed that the troubles in the DRC did not originate in tribal hatreds, but were driven by abundant resources for the picking. Legal and illegal mining fuelled most of the conflict, with mining companies and criminal exporters stirring the tribal pot in order to drive people off the land and create the kind of chaos that makes theft and low royalties easier to achieve. Thanks to globalization and pressure from the World Bank and the International Monetary Fund (IMF), many countries, including DR Congo, had liberalized mining and lowered environmental and labour codes, paving the way for degradations of every sort. According to Jamie Kneen of Mining Watch Canada, writing in 2004, "Metal prices are booming, and Canadian mining companies are taking advantage of the same prejudicial conditions to expand into all corners of the globe, manipulating, slandering, abusing, and even killing those who dare to oppose them, displacing Indigenous and non-Indigenous communities alike, supporting repressive governments and taking advantage of weak ones, and contaminating and destroying sensitive ecosystems." He cites Barrick Gold and TVI Pacific as two companies that, though they "threaten to sue their critics for libel and defamation, know that while what we are saying is true, and we have the documentation to prove it, we could never afford to mount an extensive legal defence." In 2008, Barrick Gold brought a six-million-dollar lawsuit against the small Quebec publisher Écosociété and the editors of *Noir Canada: Pillage, corruption et criminalité en Afrique*. And Banro Corporation, another company singled out in the book, upped the ante with a five-million-dollar lawsuit against the publisher and authors. Two other companies with strong Canadian links, Kinross Gold and Katanga Mining, have also been accused of dubious practices.

Although Canada is a major player, controlling or investing in more than 54 percent of mining interests in Africa and claiming or operating six hundred concessions, it is not alone in contributing to disruption, chaos and worse. In April 2001, the United Nations Security Council published the results of a three-year investigation in its *Report of the Panel of Experts on the Illegal Exploitation of Natural Resources and Other Forms of Wealth of the Democratic Republic of Congo*. According to Mining Watch, the report "exposed sophisticated networks of high-level political, military and business persons in cahoots with various rebel groups . . . intentionally fuelling the conflict in order to retain their control over the country's natural resources." Other countries involved in these activities included Uganda, Rwanda, Namibia, Angola, South Africa, the United States, Belgium, France, Germany, the United Kingdom and Switzerland; amongst the facilitating entities, the World Bank and the IMF figured prominently. A year earlier eighty-five companies had been named as contributors to violence in the DRC. Not surprisingly, a number of the key players lobbied vigorously, threatening legal action while donning a greener, friendlier mask, and were dropped from the list.

As quoted on Bloomberg.com in 2006, Clive Newall, CEO of First Quantum Minerals, a major shareholder in Anvil Mining, described the DRC as "the holy grail of the copper industry. Companies are saying: to hell with the political risk, we just have to be here." Anvil Mining was accused of providing logistical support to the Congolese army in a massacre that took place at Kilwa in October 2004. Although the company denied accusations of complicity, a military judge in the DRC recommended that charges be laid against company personnel, all of whom were eventually exonerated in what many observers consider a gross miscarriage of justice. According to GRAMA (Groupe de recherche sur les activités minières en Afrique), Banro controls 93 percent of SAKIMA, which manages 47 mining concessions covering 3,966 square miles in South Kivu and Maniema provinces in eastern DRC. The government of the DRC holds the other 7 percent. Diamonds and copper aside, the deposits in Banro's ten gold concessions alone were estimated to be about 13 million ounces, which, at a retail value of one thousand dollars an ounce, represents a string of zeroes worth fighting for.

Barrick Gold Corporation, with its head office in Toronto, had, as of 2001, former Canadian prime minister Brian Mulroney and former

U.S. president George Bush Sr. on its advisory board, along with assets of $5.45 billion. The previous year saw Barrick extract 3.74 million ounces of gold worldwide. According to Mining Watch, the company was accused by a transnational tribunal of human rights and environmental abuses in Chile, Argentina and Peru. Barrick has also been under scrutiny since its purchase of the Bulyanhulu site in Tanzania, a subsidiary of Kahama Mining, where, in 1996, under previous administration, Amnesty International claims there were evictions and extrajudicial executions over disputed land. Yet the company continues to enjoy risk insurance support from the Export Development Corporation in Canada.

War and profit are ancient comrades. Vested interests brought rebel leader Laurent Kabila to North America in the 1990s, where, it is alleged, deals involving officials in two governments and two mining companies were made to assist him in his struggle to unseat Mobutu. Once the world's major supplier of slaves, then rubber for the burgeoning auto industry, and a major source of uranium during World War II—some of which found its way by air to Hiroshima and Nagasaki—the DRC is now a major source of cassiterite, used in cellphones, computers, iPods, PlayStations and satellites. Corporate eyes are on the bottom line. Many of the miners are children, malnourished, dropping from exhaustion, forced to work underground with bare hands and no protection, some of them dying of cholera on the road as they transport huge bundles of cassiterite on their backs for the flock of small, unmarked planes ferrying the illegally extracted minerals from Walikale to Kigali, Kampala and other distribution points outside the DRC. This ongoing scramble for resources is what fuels the violence here, not anything peculiar to Africa or Africans.

The ferry reduced speed and manoeuvred into the dock in Bukavu, the capital of South Kivu province. From the quarterdeck, I could see how this charming southern capital might be a fairly reasonable approximation of Goma before the volcanic eruption of 2002. Despite its idyllic location and more prosperous air, Bukavu had a darker side. I had a UN interview scheduled for later in the day to discuss the repatriation of defecting Hutu rebels back to Rwanda and was also hoping to gather information about the illegal exportation of coltan, for which Bukavu was the hub. This crucial heat-resisting metal is extracted underground in South Kivu by slave labour, many of the miners again naked or shoeless children. I bade farewell to Célestin as we disembarked. Although I was sad to see him go, he took leave of me with all the confidence and nonchalance of youth.

Congo, Canada, the world are all part of that moving toyshop, I realized, as I fingered one of its principal playthings, the mobile phone in my pocket. Another favourite toy, one I had not yet acquired, the AK-47, was cradled in the arms of a soldier on the dock. As a privileged foreigner, and clearly implicated in the tragedies unfolding in sub-Saharan Africa, I felt I was moving deeper than politics, culture and religion, into the unexplored rifts of the psyche—which Yeats called "the foul rag-and-bone shop of the heart"—waiting, like Lake Kivu's methane gas, for the moment of their deadly turnover.

After yet another bumpy motorcycle ride, I arrived at the UN's Office of Intelligence, Propaganda and Logistics. An edgy young man who seemed to hover in his chair rattled off his responsibilities as an employee of the UN's DDRRR program in the cramped portable headquarters on the perimeters of Bukavu. He was candid about political short-sightedness and diplomatic blunders, admitting that the West had encouraged Bosnian-Serb racists to think they, too, were in the right. "We've done the same with the Rwandans," he insisted. "They are the darlings of the West. They're well funded, their officers are trained abroad, mostly in the U.K., and they now have three thousand troops stationed in Darfur, despite the fact that the deputy commander of the mission has been accused of war crimes."

I asked why Rwanda was getting such privileged treatment from foreign governments and agencies.

"It's guilt money," he told me, "for not responding to the Rwandan genocide."

I asked next about the problems in Congo, whether the colonial legacy was to be blamed, but he dismissed that notion quickly. "The Belgians are wearing thin as scapegoats," he said. "The Congolese need a new enemy, and the Rwandans fill the bill perfectly." He placed much of the blame on the DRC itself. In a census done twenty years earlier, the population was pegged at 15 million; now it was 55 million. By mid-century, it was projected to be 150 million. He shrugged. "There's no way to avert disaster, unless the thinking changes." The problem, he assured me, is that Congo invests in the past, not the future. To explain, he shared a story about a Congolese man who, faced with the decision of which member of his family to save in a life-and-death situation, chooses his mother instead of his wife or son. Why? Because you can always have another son or another wife, but you'll only ever have one mother. The parallels with Philip Winter's story about the old Sudanese man who placed more value on his culture than on children made me wonder if the constructing of parables was part of the job description for UN employees. My young informant was in favour of positive action, of nation building, he said, but did not see much incentive to move in that direction in Congo. The DRC needed a complete shakeup and retooling, though that seemed unlikely to happen. I asked him what the West could do.

"We care," he said, "but not enough." The phone rang. He rolled his eyes for my benefit, spoke briefly and promised to call back. "Look—" he held both hands in front of him, palms up, in the Buddha's teaching position. "I could have the top fifty FDLR heads in a basket within a month if I had twenty thousand dollars to offer for each of them. That's only a million dollars. The UN presence in Congo is costing us more than a billion dollars per year."

"Can I quote you by name?" I asked.

"Go ahead, I'll deny it anyway." Then he offered some free advice to the remaining Hutu *génocidaires*. If the FDLR arrived at the border and handed over their weapons en masse, they could immediately re-emerge as a serious political force in Rwanda. We were both silent for a moment, contemplating this prospect, its dangers and its unlikelihood.

"Rwanda can't afford democracy," he said, standing up and offering a farewell handshake. "The country is small, disciplined, clean and as fragile as a glass egg. One tap and it's shattered."

And Congo? For the time being, it was business as usual.

When I left the UN compound, I was frustrated and angry. I knew what I'd heard was not far off the mark. And I also knew that the 2001 statement called "Covert Action in Africa: A Smoking Gun in Washington," given by investigative journalist Keith Harmon Snow at a government forum in Washington, D.C., was still accurate in its allegations of foreign complicity in the exploitation and destabilizing of Congo. The war machine needs Congo's cobalt reserves; the electronics industry, its coltan. "Even the village idiot, if he opens his eyes, can see that the directors of the media corporations are the same directors [as] those raping Africa," Snow stated. "But too many people have a paycheck to worry about. And that includes humanitarian organizations and the United Nations and the OAU . . . What I am talking about is access. That's all. Access to the animals. Access to the game parks and trophy fishing. Access to the minerals. Access to the cheap and replenishable labor pool. Access to biological and pharmaceutical testing grounds. Access to markets. It may seem contradictory, but it's all completely unethical, entirely arrogant and racist. It is driven purely by greed. And the profound human suffering is so totally unnecessary."

I did not want to believe that the situation in the DRC was hopeless: that lava could not be removed from the runway in Goma; that the huge force of civil society, with its 103 local NGOs in eastern DRC, would have no influence on public policy; that the UN, when not encouraging mistaken alliances and abortive military engagements, was wasting time and money; that nations, my own included, would never impose and police human rights and environmental standards on their extractive industries operating abroad; that the child soldiers I'd met could not be healed and offered a chance at a different kind of life; or that Mbeda would never be free from the threat of physical or sexual violence to care for her nieces and nephews and to sew beautiful garments in Rutshuru. I wanted to believe that Célestin might become an African I.F. Stone or live to write the great Congolese novel, that Kubuya could retire with the knowledge that his political campaigns for justice had made a difference and that the small child I'd seen last fall trying to sell her tiny pyramid of potatoes amidst the volcanic rubble of Goma could grow up to have good health, education, a family and a small shop with books, clothes, foodstuffs and real toys on the shelves.

I spent the rest of the day exploring Bukavu. I'd hoped to check out various small freight and airline offices, but they had closed for the weekend, less demanding than the mines they serviced, which operated 24/7 extracting minerals illegally. Exhausted, I retired to my room at the Hôtel Résidence, a down-at-the-heels colonial edifice on the main street, with high ceilings, patterned tile floors, a covered billiards table in the upstairs lounge, a stunning view of the harbour and, alas, a throbbing, all-night discotheque in the back alley.

six *A Breath of Kings*

Several soldiers had taken up positions in front of Addis Ababa's National Hotel. This was not an unusual sight in African cities, where even the local grocery store may have an armed guard, so I did not pay much attention to their presence. I adjusted my day-pack and crossed the street to the church compound, planning to take a few photographs of the city's slums against the background of new high-rises and expensive hotels. I could make out a dozen structures of recycled tin, boards and plastic. Children scurried about like trolls in the shadow of the Finfine Bridge over the ravine. As I positioned myself, two denizens of this dreary netherworld, an old woman and a small boy, emerged from under the bridge and made their way towards the church, where a consolation of sorts awaited them. The briskness of the woman's movements belied her appearance. She might, after all, have been the child's mother. The small boy's face shone from being scrubbed, and his clean white shirt seemed intent on absorbing all the morning's brightness.

I was contemplating a descent into the ravine when a military vehicle screeched to a halt and a dozen soldiers alighted on the pavement, rapidly stationing themselves on either side of the street. When I lifted my camera to record the moment, one of them shook his head and slowly raised his rifle. I got the message. Motorcycle roadblocks had been set up at the intersection, diverting all traffic. As no vehicles were moving across the bridge, I presumed the same conditions applied farther up the hill. I was halfway across the street, on my way to the hotel, when two of the soldiers yelled at me, insisting I go back. I turned around to retreat but got the same reaction from soldiers stationed on the other side. I stopped, looked both ways and raised my hands in a questioning gesture.

A siren blared. A soldier ran out, grabbed my arm and pulled me aside just as the first of six motorcycles in a cavalcade rounded the corner from Bole Road and accelerated slowly towards us, lights flashing. A gun-mounted Jeep and two armoured vehicles, with soldiers on top covering the eight principal points of the compass, followed close behind. Five black Mercedes sedans and three stretch limos, all with heavily tinted windows, proceeded sedately past, each maintaining a car's length of distance from the one in front. Behind them came the same protective units in reverse order—armoured vehicles, Jeep and motorcycles—but with one difference. In the wake of the cavalcade, bringing up the rear in an almost comic manner, was a single white ambulance, completely out of place and a reminder that the best-laid plans, even of emperors, presidents and kings, as the poet advises, *gang aft aglay*. For all the soldiers knew, I might have been the catalyst that made everything go wrong, like the solitary figure in the crowd who shot Archduke Franz Ferdinand in Sarajevo and is said to have triggered World War I.

I thought I'd just witnessed the homecoming of Ethiopian president Girma Wolde-Giorgis, but the next day's news indicated that the special guest was none other than Libyan president Muammar Abu Minyar al-Gaddafi, arriving to recruit Ethiopian beauties as stewardesses for his national airline.

I was excited to be in Ethiopia, not only the birthplace of Lucy, but also an ancient Christian kingdom in a sea of Muslims and animists, with a lineage some trace back to Ethiopic, the great-grandson of Noah. Ethiopia boasts a royal pedigree that includes Menelik, alleged to have been the eldest son of Solomon by Makeda, the queen of Sheba, a fantastical genealogy with the even more dubious claim to have lasted uninterrupted until 1974, when His Most Puissant Majesty, Elect of God, Lion of Judah, Emperor Haile Selassie was dethroned. Axum, one of the great civilizations that flourished from the first to the seventh century, witnessed the shift from Judaism to Christianity and felt the first stirrings of Islam. Not unlike the civilization of the Maya, who held sway in the Middle Americas during roughly the same period, the land of the Axumites is steeped in legend and archaeological wonders that have stirred the imagination of explorers, treasure hunters and poets. According to archaeologist Stuart Munro-Hay,

> Britannia was only the most distant Roman province then, when Aksum, with its capital over a mile above sea level on the "roof of Africa," was listed by the prophet Mani as the third most important kingdom in a list that included Rome, Persia and China. Later a Byzantine diplomat described his audience with Kaleb of Aksum, conqueror of the Jewish king of Yemen. The embassy proposed that Aksum join the silk trade, buying from Indian merchants to exclude Rome's inveterate enemy, Persia. The ambassador witnessed King Kaleb's arrival, standing high on a dais bound with golden leaves, set on a wheeled platform drawn by four elephants. From his gold and linen headdress fluttered golden streamers. His collar, armlets, and many bracelets and rings were of gold. His kilt was also gold on linen, his chest covered with straps embroidered with pearls. He held a gilded shield and lances. Around him musicians played flutes and his nobles formed an armed guard.

Even Coleridge's famous pleasure dome is linked with this ancient land:

> In a vision once I saw:
> It was an Abyssinian maid,

And on her dulcimer she played,
Singing of Mount Abora.

Could I revive within me
Her symphony and song,
To such a deep delight 'twould win me
That with music loud and long

. . .

I would build that dome in air,
And all who heard should see them there,
And all should cry, Beware! Beware!
His flashing eyes, his floating hair!

I found the legacy of unrolling dynasties, wars, political intrigues, interventions and occupation—thanks to Arab, Egyptian, British, Italian, Soviet and American meddling—both dazzling and dizzying. And I was unsure how to conduct myself here. The Ethiopian government, jittery because it had won only a small majority in the last election, and not one of the twenty-three seats in the capital, had introduced draconian laws and measures to restrict the movement of people and the influence of foreigners. The subject of human rights was now forbidden, and talk of displacement and famine could result in arrest. I was amazed, as a *faranji*, a foreigner, that I had not already been searched or questioned. Once the soldiers and crowd dispersed, I retrieved my hat from the hotel room, to corral my own floating white hair, and sunglasses to protect against the solar intensity of the Ethiopian midday.

I had arrived in Ethiopia with enough disturbing stories to last several lifetimes and was not anxious to add to that storehouse of pain and suffering. My plan was to interview NGOs, academics, government officials and ordinary citizens to see what I could learn about justice, humanitarian aid and the situation of refugees. I wanted to know whether the huge sums of money from donor nations and organizations were promoting or impeding development and what the effects were on human rights. My friend Gerry Caplan, author of *The Betrayal of Africa*, a moving indictment of the smugness and ruthlessness of Western business and political interests, was convinced that "we in the West are deeply complicit in every crisis bedevilling Africa, that we're up to our collective necks in retrograde practices, and that we've been virtually co-conspirators with certain African leaders in under-developing the continent and betraying the people." He had recommended that I talk to his friend Wolde Selassie, an authority on questions of foreign aid.

"Wolde is a gold mine of information and someone who's dedicated to change and healing. Don't start anywhere else."

Wolde and I arranged to meet for coffee at the National Hotel. He had completed a doctorate in Germany, he informed me, but while many of his Ethiopian classmates stayed in Europe or took up lucrative teaching jobs in the U.S., he chose to return home out of a sense of commitment to the communities he'd studied for his degree. He had analyzed a relocation project, widely referred to as "villagization," in which 82,000 people from the Ethiopian Highlands had been shifted to a more fertile region in the south. After a few years, only 26,000 of the original group remained there, though they had been well provided with the basic human needs of food, shelter and clothing. What, Wolde wanted to know, had made them return to the highlands? He began to understand the role of place in culture, the importance of something as simple as wanting to be where you were born, where you knew every inch of terrain like the back of your hand, where your social position in the network or hierarchy of responsibility was fully understood. He rejected the assumption, he said, that humans are primarily profit-making machines rather than complicated moral beings whose antennae are attuned as much to climate, weather and patterns of migration as to social, political and economic variables.

Wolde had chosen to work for various agencies rather than accept the teaching positions for which he was so eminently qualified. To introduce the subject of aid, I mentioned my encounter with Victor Ochen at African Youth Initiative Network in Gulu and his story of how a sack of rice in the right hands could achieve so much. Wolde nodded and offered me a local example. A modest budget of 100,000 Ethiopian birr (ten thousand dollars) from a donor was made available to a community in the southern region of Ethiopia. The community leaders met and decided to give 10,000 birr to each of ten groups, including a youth group, a burial society, a Christian congregation and a Muslim organization. The youth group, in turn, gave each of its ten members one thousand birr. When Wolde returned, he saw that the ten youths had used the money to start a shoeshine business and a small enterprise selling gum. The church group (with an additional 2 percent contributed by the minister) used its money to start a candle-making business. The moral of the story was obvious: the members of a community know best what they need.

Wolde took a sip of his coffee. "The opposite approach is far too common," he said, and gave two examples. First, the case of a foreign NGO with an operating budget of $150,000, in which all of the money went to pay for rent, salary and a huge Land Rover for its single employee; not a penny remained for projects on the ground. Second, an outrageous sum of $290 million from Italian aid for Italian engineers and construction workers to build a dam that silted over, a rice processing plant that now lay in ruins and a facility for manufacturing irrigation pipes that was now defunct. No locals had been consulted or hired. The only consolation, Wolde said, since the Italians had inherited one special skill from their Roman forebears, was that the road they built was still in use. Like much so-called foreign aid, the money itself ended up back in the coffers of the donor country's companies and entrepreneurs.

I could see why Wolde and Gerry Caplan were mutual admirers. According to Gerry, "So long as Western countries treat aid as a political tool to advance their own self-interest, and so long as most international NGOs compete against one another, the prospect of a more rational and less wasteful system remains a pipedream. In the meantime, we criticize Africans for being inefficient."

I could see also how the practice most NGOs had of paying local workers a tiny fraction of the salaries paid to expatriates could be a major source of irritation for someone with Wolde's knowledge and training. He would be expected to take advice from people with less education, no local knowledge and, most often, little or no experience in the field. For many expatriates, these postings were merely a stepping stone to something better. But Wolde was not bitter or cynical; in fact, he was positive, generous and outgoing. It was encouraging to know that he had not given up on those in need or on the aid industry itself.

ETHIOPIA REMAINED A largely feudal society long after Haile Selassie returned to the imperial throne at the end of World War II, a situation that, given rising expectations and the impact of African liberation movements in the 1960s and 1970s, could not last. Most Ethiopians—herders and subsistence farmers—lived or died according to the vagaries of nature, and their country's complete lack of infrastructure, as well as a serious shortage of industry, made change inevitable. When Haile Selassie was overthrown in 1974 by the Derg, which began as a left-leaning military coordinating committee, the country lurched into seventeen tumultuous years of ruthless suppression, misguided collectivization, famine and civil war. Eritrean liberation movements fought to regain that country's independence, and Ethiopian opposition to the ruling Derg escalated in every region. The political situation during this period was unique: a socialist regime receiving arms from the Russians but forced to accept humanitarian aid from the United States. When the Soviet Union collapsed, the Derg lost its principal military supply line and was soon defeated by the combined forces of the Eritrean People's Liberation Front and the Ethiopian People's Revolutionary Democratic Movement (EPRDM), a title coined to achieve legitimacy in the eyes of the world, East and West.

Although Eritrea achieved independence in 1993, its love affair with EPRDM did not last. A currency change and border skirmishes over the disputed Yirga Triangle led to major hostilities, with the expulsion of foreign nationals and enormous casualties on both sides. The dispute was not resolved until 2000 when, under pressure from the international community, Eritrea and Ethiopia signed a peace agreement. However, when I arrived in Addis, borders remained closed between the two neighbours.

After spending years covering the Ethiopian-Eritrean wars and the deteriorating situation in Sudan, American journalist Robert D. Kaplan, in *Surrender or Starve: The Wars behind the Famine*, claims the notion that "western aid in the long run could kill more people than it saved in the short run was neither farfetched nor unfair." He exposes the frequent folly of good intentions and insists that close analysis of these conflicts "should cure those in the West of the delusion that humanitarian means are sufficient to achieve humanitarian ends in Africa." In Kaplan's view, acts of God have been less significant than power relationships in the tragedies unfolding in Africa: "It almost doesn't matter that in the process of saving millions in Ethiopia, the West may have salvaged Africa's most chillingly brutal regime, thereby giving it the wherewithal to ruin the lives of even more millions during a long period of time. Graphic images of starving children simply couldn't be ignored."

Shortly after arriving in Addis, I interviewed Kunal Dhar, a Canadian working for the Department for International Development (DFID), a U.K. aid agency deeply committed to Ethiopia. As we sat in the beautiful sun-filled offices in the British embassy compound, he offered a sobering round of statistics: 81 percent of Ethiopians live below the poverty line; 90 percent of the country's budget consists of foreign aid; 100 million dollars in aid was provided in the last four months of 2008; 673 women die in childbirth out of every 100,000; and 123 children out of every thousand die before age five. I was not ready for this information; even less so, for the numbing list of acronyms that kept popping up in his account, each of which had to be spelled out for me: MOU (memorandum of understanding); WASH (water, sanitation, hygiene); PSN (productive safety net); and MDG (Millennium Development Goals). The one that caught my attention was RBA (rights-based approach), an attempt to tie continued aid to evidence of human rights reform.

"Who decides if an organization or its work is accountable?" I asked.

"Accountable to whom?" Kunal redirected the question back to me. "To donors? To the people who are being served? To national governments?"

"Why not all three?" I countered. This controversial issue, which sounds simple on the surface, was causing a lot of havoc amongst NGOs, whose ways of operating often required playing by several different sets of rules. I'd been reading books on the subject, including Peter Uvin's *Aiding Violence*, in which he argues that "all development aid constitutes a form of political intervention . . . at all levels, from the central government to the local community. Ethnic and political amnesia does not make development aid and the processes it sets in motion apolitical; it just renders these processes invisible." I mentioned this to Kunal, along with Uvin's specific references to Rwanda, where he claims aid contributed, unwittingly, to the genocide: "On the level of practice and discourse, the aid system did not care unduly about political and social trends in the country, not even if they involved government-sponsored, racist attacks against Tutsis, many of whom were aid employees or partners."

"I approve of the Paris Declaration on Aid Effectiveness, which seeks to harmonize and manage aid for results that can be monitored," Kunal explained, offering me tea. "But I have reservations about its usefulness and applicability."

I mentioned Naomi Klein's *The Shock Doctrine*, which had impressed me for its detailed and passionate denunciation of the Chicago School of economists, the head priests of laissez-faire and the deadly structural adjustment programs imposed on defaulting countries, which forced the elimination of free public education and health care.

Kunal nodded. "Very bright, very committed. We were university classmates in Ottawa."

"So much for the invisible hand of the marketplace."

Kunal laughed. "Yes, it's more like an iron fist." Conversation turned to *Confessions of an Economic Hit Man: How the U.S. Uses Globalization to Cheat Poor Countries Out of Trillions* by John Perkins. Had I read the book?

My turn to nod. Perkins had been a CIA operative whose cover was that of a chief economist. His job was to encourage developing countries to borrow vast sums of money they would never be able to repay, thus rendering them vassals of U.S. foreign policy. Ninety percent of those loans went directly to American companies, such as Bechtel and Halliburton, which built the new infrastructure that would benefit only the leader and his closest associates. "What it is all about," Perkins says, "is building empire . . . and in this process, we have been able to build the largest empire in the world." If the economic hit men failed, he explained, the "jackals" (CIA-orchestrated operatives) were sent in to finish the job; that is, to foment revolutions, fund coups or arrange assassinations, including those of Chile's Salvador Allende, Panama's Omar Torrijos and Guatemala's Jacobo Arbenz. Five percent of the world's population uses and controls 24 percent of the world's wealth and resources; 24,000 children die of starvation every day; a similar number succumb to curable diseases. This, Perkins admits, makes a lot of people angry, especially spiritual communities like the Muslims who see their world being pillaged. His comments reminded me of an observation by Uruguayan writer Eduardo Galeano: "The big bankers of the world, who practice the terrorism of money, are more powerful than kings and field marshals, even more than the Pope of Rome himself. They never dirty their hands. They kill no one: they limit themselves to applauding the show."

Before I left, Kunal gave me a list of contacts in Addis. One of these names, Jerusalem Bernahu, was also on the list I'd brought with me from home. This seemed a good omen.

"Conditionality is a dirty word," Jerusalem said the next morning when I visited her office adjacent to the Canadian embassy, where she worked for CIDA, the Canadian International Development Agency. She also participated in a donors' coordination subgroup on human rights and development. Although she was professional enough to temper her comments, I could see she was well informed and sensed her frustration with political stonewalling at the local level. Ethiopia, like Sudan and an increasing number of African countries, preferred China's modus operandi: here's the money, no strings attached. "No ethical considerations, maybe, but plenty of conditions in terms of trade."

"Governance is a problem," she added. And with this colossal understatement, Jerusalem launched into her own detailed description of Ethiopia's economic disparities and political rigidity. "It's a very poor country, with a plethora of beggars, the disabled and the unemployed, but these are symptoms, not causes, of the dysfunction." I was staying downtown, she noted, so I'd obviously seen the slums, the open sewers and garbage dumps where children and goats competed for food scraps in ravines alongside expensive hotels and spanking new apartment buildings.

"The wealthy, who kept their money safely abroad during the Marxist period of the Derg, are now beginning to invest in Ethiopia, but the pace of development in the arena of social services and infrastructure is abysmally slow."

So many people doing good work, yet so much money being wasted on the wrong projects or winding up in the wrong hands. Another woman I interviewed later that afternoon, an employee at the Swedish embassy, would give me another perspective on the problem of aid. She was not only interested in the work I was doing, but also touched by the conditions I described in Congo and Uganda. Her eyes welled up with tears as I related the story of Nancy's mutilations by the LRA.

"Ethiopia produces enough food to be self-sufficient," she said, "but between seven and eight million people are still going hungry. Explain that."

"Why are you doing this kind of work?" I asked. "What keeps you going when conditions here seem so hopeless?"

Some of her friends, she said, had gone into the oil business; others had stayed home in Sweden and become successful professionals. "I know this will continue whether I'm here or not, but I want to be on the right side of the problem." It was then she advised me to get out of Addis if I wanted to know the real story.

GETTING OUT OF the city was not as simple as I'd expected. Flights were limited and rural access by writers and journalists was closely monitored by the government. Besides, I had commitments, including a lecture I had promised to deliver at the University of Addis Ababa.

"We have lots of sun here, but no electricity," Professor Bernahu Matthews explained as we walked across the campus.

He apologized for the communication problems we'd had—lost and unanswered e-mails—which he attributed to inadequate and poorly maintained facilities at the university: phones, faxes and computers rendered useless whenever the electricity went down. "The government is talking about supplying electricity to Sudan, Djibouti and neighbouring countries when its own citizens do not have an adequate supply." A few backup generators on campus might ease the problem, I suggested. And what about green alternatives? Bernahu agreed that the capital, with its exposure and elevation, would be an ideal location for producing wind and solar power. As we walked to his office, he took a perverse delight in pointing out the architectural anomalies on campus: the elegant palace and parklike setting Emperor Haile Selassie had donated to the university, the dreary Soviet architecture constructed during the period of the Derg and the ultramodern Kennedy Library, a post-Soviet gift from the United States.

A graduate in linguistics from the University of London, Bernahu had developed a passion for oral literature and folklore. "My language of birth is not Amharic, but Oromiffa, spoken by the south-central Oromo people, the largest ethnic group in Ethiopia." He looked up at me as he jiggled the key in his office door. "Amharic is spoken mainly by politicians from the north; that's why it has become the national language."

The Oromo peoples in Ethiopia have a long history of persecution, including murder, harassment and arrest. Two of their most famous singers, who have achieved the status of international pop stars, now live in Canada. Kemer Yousef, twenty at the time, was part of a group of two hundred men, women and children who fled on foot to Somalia more than twenty years ago. According to journalist John Goddard, only thirty-seven survived the bandits, snakes and other hardships. Ali Birra, another Oromo singer, left Ethiopia in 1993.

I was relieved to find that Bernahu had an independent mind and a good sense of humour. In order to establish contact with him and his colleagues, I'd offered to give a lecture about literature and politics. However, knowing that writers, singers and opposition politicians had recently been jailed, I also proposed an alternative: the links between poetry and history. As Bernahu seemed untroubled by my preferred topic, I suggested "Killing the General with Words: Writing as a Subversive Activity" as the title for my talk. Without the slightest sign of hesitation, he dutifully copied this down and requested some biographical information to include on the poster, which would appear on the bulletin board by the door to the cafeteria. A date was set for the following Monday afternoon.

I made my way back to the main entrance to the university, past the library and through the imposing stone gates. As I emerged onto the busy street, hundreds of students were rushing to and fro or milling about in small groups. A dozen mini-buses spewing diesel fumes jockeyed for position while their touts shouted to attract customers. In the midst of this mayhem, a middle-aged man sporting only a beard and a wry expression sauntered along the sidewalk, trousers in one hand, his naked pot-belly and shaggy genitals attracting little attention from passersby.

My base in Addis was the National Hotel, an inexpensive, second-rate high-rise run by the government. The lobby, with its cheap plastic furniture and mounted trophies of horned animals, looked like an unfortunate marriage of bus terminal and hunting lodge. The National was also an architectural hodgepodge, with two huge concrete pillars in front. One of these ran through the centre of my room, as if hurled there by an angry Titan. The main problem with my lodgings, in addition to mould, faulty wiring and electrical blackouts—the hotel, like the university, did not have a working generator—was the very real possibility of spending a day trapped in the elevator. But I liked the price, the modest aspect of the place and its friendly, no-nonsense staff, though the noise from the church across the street, with its loudspeakers cranked up to full volume, was driving me crazy.

Kidus Istephanos Ethiopian Orthodox Church was a microcosm of the country: ornate and Byzantine, but also crass and commercial. As lacklustre in appearance as the National Hotel, the church had somehow acquired a generator. Either that, or it was plugged into a demonic power source. Blackout or not, high-pitched, abrasive calls to worship and crude spiritual ranting rang out uninterrupted from morning to night. No need for a John the Baptist to announce your arrival when you've got an amplification system like this. And why use political torture when religious music and rhetoric can do the trick perfectly well. Two pillows over my head did nothing to diminish the cacophony. Although the church proper was closed to the public, who were obliged to worship outside, regardless of the weather, occasionally a flesh-and-blood priest would pick up an external microphone to address the motley gathering. Beggars, the lame, sellers of candles and incense lined the concrete stairs leading up to the church, along with dwindling numbers of the faithful crossing themselves and hoping for better luck in the afterlife. Large advertisements for Coca-Cola appeared at either end of the church compound—one a red equipment shed with the company logo printed in large white letters, the other a twelve-foot concrete Coke bottle that had seen better days, its paint flaking—completing the picture of an institution that had sold out to the marketplace and was trying desperately to make up in decibels what it had lost in spirit. According to newspapers, Coca-Cola's branch plant had closed for an unspecified period in response to the economic downturn. I preferred to think the closure was a result of the growing sophistication amongst the Ethiopians. With any luck, their addiction to this toxic opiate would prove short-lived.

The current regime under Wolde-Giorgis had come to power in 1991 straight from the bush, with very little education or political know-how, only to find itself at war with a fiercely secessionist element in its midst. Infrastructure and social services played second fiddle to military buildup and expenditure. Although human rights and civil liberties were currently on the shelf, the people I'd spoken to were hopeful that this was temporary. Superficially, the city seemed tranquil, with only a modest police presence. I'd been told that Eve Ensler's play *The Vagina Monologues* was being staged downtown, and my lecture on literature and politics was scheduled to go ahead with no apparent reaction from the authorities. However, the ongoing harassment of those in political opposition and an insidious network of neighbours spying on neighbours suggested a different reality. The arrest and incarceration of pop icon Teddy Afro on trumped-up charges of manslaughter and the new regulations that restricted local NGOs to only 10 percent foreign aid indicated the government was running scared and determined to do whatever it deemed necessary to hold onto power. The most chilling news was the re-arrest and sentencing of Birtukan Midekssa. A lawyer, former judge, single mom, champion of human rights and charismatic leader of the biggest opposition party in Ethiopia—a woman who has been compared with Burma's Aung San Suu Kyi—Midekssa had been jailed after declaring the 2005 elections a fraud. International pressure prompted her eventual release, but she was re-arrested on December 28, 2008, and sentenced to life for giving a speech in Europe about her illegal incarceration.

Awakened far too early by the electronic rumblings of the Orthodevils across the street, I decided to walk up Bole Road to an Internet café and wait there for the Southern Sudan mission office to open. Sudan was notoriously difficult to enter. You needed an official invitation or some sort of connection. And the political situation had heated up considerably since March 4, 2009, when the ICC issued a warrant for the arrest of Sudanese president Omar al-Bashir for war crimes and crimes against humanity. He defied the announcement publicly, made official visits to neighbouring Arab countries and expelled a number of foreign aid organizations, including those trying to ameliorate the crisis in Darfur. Visas to Sudan were no longer available, though Southern Sudan was rumoured to be open for business through its own mission in Addis Ababa, issuing alternate entry permits. Three weekly

Ethiopian Airlines flights between Addis and Juba were still operating. After watching the Internet connection disappear on my computer screen as I tried to check e-mail, I set out down the labyrinthine back streets, past India House and other official establishments. It was a beautiful morning, sunny with a light breeze. Even the stray dogs smiled at me. However, when I arrived at the mission compound, a sign on the door announced that the office would be closed all week and no visa applications would be processed.

I stood there with my mouth open while the truth sank in: I would not get to Southern Sudan, a place I'd been preparing to visit for months. I had applied six weeks earlier. Although none of my e-mails to the mission were answered, I'd been assured by contacts in Juba that there would be no problem acquiring a visa. As an election on independence was scheduled for 2011, I expected to talk to many individuals who had fought and prepared the ground for this new possibility. The midday sun showed no mercy. I considered knocking on the gate to see if I could rouse a sympathetic response, but realized the futility. I would only irritate the security guard asleep at his post inside. I kicked a loose stone across the dirt road. One of the stray dogs started barking, its erstwhile benevolence morphing into a nasty snarl. Okay, I thought, all the more reason to get into the field in Ethiopia.

I made a quick decision and phoned a contact working in Addis for a major international NGO. He agreed to meet me at the Imperial Hotel. For reasons that will become clear, I'll call him X and will not name his NGO. Over coffee, he acquainted me with its three main pillars of support—nutrition, education and health—and told me it was currently assisting between seven and eight million IDPs and refugees in Ethiopia. Some emergency work, including a flying health unit and food drops, was undertaken to respond to crises. The organization had a lot of complicated projects in the Ogaden region, he said, the largely Somali territory that Britain had ceded to Ethiopia in 1934, a decision still very much in dispute. The situation there was desperate, thanks to rebels and the collapse of civil society in neighbouring Somalia.

Working in Ethiopia was not easy, he advised me; in a modern, multi-ethnic state, everything required an enormous amount of consultation and negotiation. When I asked how the impact of the rights-based approach to aid was playing out in Ethiopia, he insisted that the work of his NGO was not ideological; its task was relief, not advocacy. I wondered how he could say this and keep a straight face. I mentioned the notorious new law initiated by the Ethiopian government that restricted local NGOs to only 10 percent foreign aid.

"It's hypocritical," I said. "While its own budget consists of 90 percent foreign aid, the Ethiopian government restricts local NGOs to 10 percent." The government was blaming its poor showing in the 2005 election on dangerous agendas such as human rights, being propagated in the country by foreigners. "These local NGOs will have their hands tied."

"Yes," X admitted, "the law will eliminate many local NGOs and reduce the rest to the mundane task of the building of latrines and boreholes."

The larger foreign NGOs, including Save the Children France, USAID and the U.K.'s Department for International Development, were exempt from these strictures, though they, too, had been warned to avoid overt advocacy work. The situation of NGOs reminded me of the six hobbled donkeys I'd seen earlier that day making their way along the sidewalk on Bole Road with ridiculously tiny steps.

Although my contact and I shared some political views and experience in Latin America, it did not surprise me when, later in the day, I received an e-mail from his director of operations turning down my request to visit the group's field operations. The letter was brief and to the point, but slightly open-ended:

> Given the sensitivities here regarding journalists, media and any type of report, the events in Sudan, NGOs thrown out of country on the weakest of excuses but citing articles written by the organisation or about it . . . we are cautious about hosting journalists

etc without 1) knowing for what publication and in what vein the piece will be written, and 2) whether we need to seek govt permission . . .

If you would like more assistance from us can you answer the issue of which publication, and type of piece (paper, book, academic article etc) and from what angle you wish to write. Do you have any accreditation with the Min of Information or other ministry here? Have you arranged your own transport / interpreter?

The recent HR report from the State Department was dismissed by govt here as a fabrication drawn up by a couple of people based in an office in Washington and not at all factual. Hence it's not a conducive time to be researching HR issues and not a good time for us to be supporting it. Sorry.

But please reply and we can take it from there.

Regards

Disappointed, but still hopeful, I sent an immediate reply. I was aware of the difficulties, I wrote, and would certainly proceed with caution. I hoped my book would bring conditions in sub-Saharan Africa to the attention of people in the English-speaking world by focusing on one individual's struggle to navigate the labyrinths of foreign aid and intervention. My letter prompted a second response from the director of operations, clarifying his organization's position and suggesting some alternative approaches:

Having spoken to field staff we feel that it is not in our best interests to be seen supporting a piece that has the central theme of Human Rights. Whilst we have a stated mission to support child rights, in this context, and especially with the introduction of the Civil Society organization law that was passed earlier this year with an article explicitly denying international organizations to deal directly in child rights, we have to find more technical, less vocal ways of delivering them.

> We are sorry we can't help you directly as we support the work you are doing.
>
> As a substitute we can advise that if you fly to Lalibella you will easily find transport and English speaking guides who can assist you to get out into the community. The areas around Lalibella are suffering from last and this year's drought and are showing both high levels of malnutrition (we are intervening in some nearby *woredas*) and the need for relief interventions which are ongoing . . . Those areas are also traditionally prone to issues of early marriage and other harmful traditional practices.
>
> An hour's drive from Lalibella in any direction will lead you to countless communities where you can easily talk to the people, and on a one to one basis through interpreters, and get a feeling of the traditional and current context.

It was excellent advice, but another day had come and gone, and the likelihood of booking a last-minute flight to Lalibella and making the contacts I'd need seemed remote. Several other individuals I'd counted on to help me had cancelled at the last minute because of emergencies in the field, so I called Laura Buffoni at the United Nations High Commission for Refugees. She agreed to meet me for lunch at the Lime Tree Restaurant on Bole Road.

"There are 40,000 Somali refugees in Addis Ababa," Laura said, "and 800,000 refugees in the whole of Ethiopia, from Sudan, Eritrea, Rwanda, Congo, Somalia, even as far away as Burundi. You should have no trouble making contact with them; in fact, they're all around you on the streets, many unregistered, without documents, and reduced to begging." I knew what she meant. Every time a taxi stopped at an intersection, children, cripples, the blind, women with infants in their arms appeared at the windows asking for money to buy food. An offering to one of them would instantly triple the number of beggars, fights erupting over access.

"Where do I begin? I can't start on the street." It was a pathetic comment, I realized, but the sentiment was genuine. I needed some help finding a translator and people who could articulate their situation.

Laura took out a notebook and copied down two names and some numbers. "Here's the contact information for Mulugeta, who works at the Jesuit Refugee Service, and for Yusuf, a leader in the Somali community." She stashed pen and notebook in her small bag and stood up to go.

"Good luck," she said. "You can't go wrong with them." If these names turned out to be as important as I hoped, Laura would deserve more than the huge book of sonnets the Italian poet Petrarch had written to his muse of the same name. Mine would deserve, at the very least, a series of novels, and whacking big ones at that.

THE JESUIT REFUGEE Service was an international organization with centres in many parts of the world, but I'd never have guessed that on entering the compound in Addis. Aside from a few small offices, meeting spaces and a room containing a dozen computers, JRS was very low-key. The grounds of less than an acre were unpaved; even the shed that served as a canteen had a dirt floor. Two dozen refugees were milling about when I arrived, several of the men taking shots at the single basketball hoop. Mulugeta, the project director, gave me a brief tour. He had not only come highly recommended by Laura, but also had a degree in sociology and social anthropology from Addis Ababa University and a diploma in software technology. Over the next few days, I would come to admire him and his work among the refugees. He introduced me to a young refugee named Ruboya Sean Seith.

Ruboya, who was in his mid-twenties, spoke English, French, Swahili and Kinyamulenge. While we sipped sweet Ethiopian coffee at the canteen, he shared his story with me. To my surprise, he was from Bukavu, the South Kivu capital I had just left in DRC. I asked him what had brought him to Addis.

"My parents were killed by the joint forces of Mai-Mai militia and government soldiers in 2005," he said. "We had to leave. There was no other choice." I thought of the girl I'd interviewed in Goma, who'd joined the Mai-Mai as a child, curious to know how to use a gun. Ruboya and his brother fled over the border into Uganda, where they were arrested immediately, tortured and accused of being spies. For days he urinated blood from ruptured kidneys. Once released, the young men hit the road again, hitching a ride on a banana truck to Kampala. "We had no money or documents. The churches were all closed, but we found refuge in a mosque and were given work and lodging in a Muslim household for a month. I worked for them as a servant and my brother helped out in a garage and tire shop."

This arrangement did not last. After bouncing back and forth between Kampala and Nakivara refugee camp, the brothers decided to head to Nairobi. The same ethnic tensions that had driven them out of DRC were waiting for them in Kenya.

"Don't go to Kakuma camp unless you want to die," the bus driver had warned them. "Life is cheap there." In Ethiopia, Ruboya had no work, no hope for the future. He hung out at the Jesuit Refugee Service to play basketball. "Not good," he sighed. "My brother is depressed and never leaves the room."

By now, there were about forty refugees of all ages in the JRS compound, standing in groups, clustered around the door of the computer room or jockeying for position beneath the basketball hoop. I had barely left the canteen when Chang Ter Pusch approached me. Chang, who looked like a rap poet, wore a toque, dark glasses and a red scarf. He was twenty-four, he told me, and from Sudan, where his father, who died in 1993, had been a plumber and a fisherman. When hostilities separated him from his mother and sisters, Chang ended up in UNHCR's Dima refugee camp. While collecting firewood, he had regular altercations with Ethiopian rebels and was afraid of being conscripted by the Sudan People's Liberation Army, which used the refugee camps as recruiting grounds for underage boys. In Addis, he was unable to work, because it was illegal for refugees to take paid employment that could be done by Ethiopians. Chang told me he wished to become a doctor. However, he had eye problems, a cloudiness that sounded very much like scotoma, a blind spot often caused by a ruptured blood vessel in the retina. I'd had this problem myself, and it had required laser surgery.

Chang was in contact with a Sudanese organization called Cry for Help, but his own cries had gone unheard. No opportunities had opened up for him. Despite his eye problems, he spent most of his spare time reading.

"You can lose your shoes, even your eyesight," he said, "but knowledge can never be taken away."

Mulugeta emerged from the office with a message: Laura Buffoni had just phoned from UNHCR to say her colleague had arranged for an Eritrean refugee named Ephrem to meet me. "He's special," she had added as an afterthought. When he located me in the compound, Ephrem was well turned out, spoke excellent English and, as I soon learned, had a degree in psychology from Asmara University. He was thirty years old, born in 1978 in Mendefera, southern Eritrea, and had begun his post-secondary studies at Addis Ababa University. In June 1998, shortly after hostilities broke out between Eritrean rebels and the Ethiopian army, Ephrem was dragged from his dormitory room at 5 AM by soldiers. Suspected of being a spy, he was fingerprinted and detained without charges in a concentration camp, with no blankets, little food and no freedom of movement, not even to use the latrine.

As we squatted on the hand-hewn benches in the canteen, Ephrem recalled grimly that the International Committee of the Red Cross had delivered only useless items such as table tennis equipment and volleyballs. It seemed more likely to me that the guards had confiscated the essentials for their own use. On more than one occasion the prisoners and facilities were tidied up and presented to foreign visitors for propaganda purposes. Just as quickly, conditions became intolerable again. All detainees suffered from diarrhea as a result of contaminated water, he said, and two of his friends died from medical complications. After eight weeks of being brainwashed and shifted from camp to camp, he was deported to Eritrea.

It was getting late in the day, so I suggested he join me for dinner that evening at the National Hotel. I needed time to unwind and think about what he'd told me. I'd felt a strange disconnect between the horror of his story and the matter-of-fact manner in which it was recounted. He was, after all, a psychologist, aware of his own reactions and capable of keeping them under control, whatever reason he might find for doing so. When he arrived at the hotel, the lights had just come back on, though the kitchen in the lower-level restaurant had not quite swung into action. The waiter came over to our table.

"I'll have a beer," I said. Ephrem ordered a soft drink. In his sports jacket and open-necked shirt, he could easily have passed himself off as a college tutor. He flashed me a break-the-bank smile and held up his glass for a toast. "To your health."

"What happened back in Eritrea? I gather you were able to continue your education."

"It was not promising at first. I was monitored closely. But then they allowed me to complete my degree." The good times did not last long. He was required by the government to do unpaid agricultural labour in the summers. When he complained about the conditions and lack of adequate food, Ephrem fell afoul of the authorities. "Being a Christian and having a mind of my own did not help. I was a convenient target for persecution."

He finally graduated and, with several strikes against him, it might have proven a hollow victory. However, Ephrem managed to make a career for himself by doing psychiatric social work and giving educational seminars and radio broadcasts on health matters. But this period of challenging work would come to an abrupt end. "Because of my years of study in Addis, they accused me of being a spy," he laughed. "My own country!" Accused of supporting the outlawed political opposition in Eritrea, known as G-15, he was transferred without consultation from the Ministry of Health to the Ministry of Education, where he was recruited for ideological training. "I couldn't do that, not after what I'd been through." When he refused to work for Eritrea's intelligence service and spoke out on behalf of incarcerated student leaders, two of whom had died in captivity, he was branded a traitor and arrested. No charges were laid. Two months later, he was released and ordered to stay silent about what had happened.

We ordered chicken, vegetables and *injera*, the rubbery, crepe-like flatbread that is an Ethiopian staple, useful as a vehicle for transporting food to the mouth.

"I haven't learned to eat with my hands," I apologized, reaching for my fork. By way of instruction, Ephrem tore off a strip of *injera* and placed it over a piece of chicken, using his thumb and fingers to secure the load. "Sure," I said, " it looks easy enough when you do it." I stuck with my fork, more anxious to hear the rest of Ephrem's story than to practise local eating habits.

"An offer came for me to do graduate studies in South Africa," he continued, "something I'd dreamed about. But the government refused to let me travel." A seminar he organized on healing and forgiveness, at which candles were lit in commemoration of lost friends, landed him once again in trouble. "By now," he said, "I had taken to spending most of my time at home." In the middle of the night, two plainclothes policemen came to the house. When he resisted arrest, he was beaten, his brothers threatened with a gun, and he was whisked away to a series of prisons and lock-ups in the mountains, first in galvanized metal containers, later in stone cells, abused, hands tied behind his back, and without shoes or proper food. Execution threats came daily.

Ephrem had touched almost nothing on his plate since showing me how to employ the *injera*. My own plate was only half empty. The waiter hovered nearby, concerned the meal had been unsatisfactory. Ephrem stopped talking and used his napkin to wipe his forehead. He said something in Amharic to the waiter, who retreated to the kitchen and returned a few minutes later with bottled water for both of us.

During a sandstorm, Ephrem told me, he had managed to escape and locate friends who helped him cross the border. When he requested asylum in Ethiopia, however, interrogation and detention followed. He was finally released into the hands of UNHCR. He had achieved official refugee status, but it was a moral and professional cul-de-sac that left him ostensibly free but forbidden to accept paid work and, therefore, unable to enrol in further studies.

"That's about it," he said, "the abbreviated version." Only the sound of water running in the kitchen could be heard. Ephrem leaned towards me across the table and whispered, "To tell you the truth, I'm desperate for help." The rest of the story came tumbling out. How he lived in fear of being kidnapped by Eritrean spies and informers. How he was plagued with guilt over the unknown fate of his brothers and the dangers his exile had created for his family in Asmara. And how, as a trained psychologist and professional health care worker, his mental condition and jobless state were sources of grief and humiliation.

I didn't need to be a professional to recognize the signs of post-traumatic stress disorder.

SUFFERING FROM INFORMATION overload, I slipped away the next morning to spend a couple of hours at the National Museum, situated near the university. Although another electrical blackout had occurred, the museum remained open to visitors. I groped my way from room to room, squinting at exhibits. Deposed emperor Haile Selassie's throne, which had been removed during the Italian occupation of Ethiopia during World War II and returned in 1972, was, ironically, the only artifact receiving enough natural light. It was a vertical rectangle made of intricately carved hardwood, inlaid with gold and ivory. Above the red velvet seat and backrest extended a canopy mounted with a carved wooden crown flanked by two urns. On the second floor, I studied nude sculptures and half a dozen paintings in the gathering gloom. One dynamic canvas, whose vibrant reds had managed to absorb the last scraps of light from the outside, was called *Genital Mutilation.* In another, a robed figure on horseback turned out to be St. Mercury, a surprise to me and, I suspect, to most hagiographers.

I considered the key attractions of Ethiopia to be its archaeology, including underground churches carved out of rock, and the discoveries that had caused the country to be called the Cradle of Mankind. This was the birthplace of Lucy, long considered the oldest and most fully articulated skeleton of a hominid, a mere 3.2 million years old. Her remains were found by Dr. Donald Johanson on November 24, 1974, near the Awash River in Ethiopia, a year after his chance discovery of a locking knee joint, an adaptation that enables humans to walk upright. I longed to visit the Rift Valley, but had restricted myself on this trip to human rights interviews and book-related research. No tourism. I had been destined, however, to meet Lucy earlier, and closer to home, when she was on tour in Seattle, less than a hundred miles from Victoria. In a controversial move, Lucy had been released from her storage vault at the Ethiopian Natural History Museum and permitted to travel to the Americas as part of a major touring exhibition—almost as far as her bipedal, meat-eating counterparts and their offspring had progressed in the continuing waves of human migration that swept them from Africa to every corner of the earth.

A member of the species *Australopithecus afarensis*, and only three feet, eight inches tall, Lucy owed her Western name to the fact that the group of young anthropologists were Beatles fans and had been playing a tape of "Lucy in the Sky with Diamonds" repeatedly in camp the evening of her discovery. Her name in Amharic is Dinkenish, which means "You are beautiful (or wonderful)." The seventy-odd reassembled fragments of bone laid out on dark velvet in a glass cabinet in the Seattle exhibit did have a certain beauty, especially the arrangement of ribs, which resembled those many-layered African necklaces made of amber. But her jaw fragment and pieces of skull might well have been arranged by Picasso, they left so much to the imagination.

In the end, it was neither Lucy nor Ethiopia's rich heritage that preoccupied me, but Ephrem's story and the stories of so many other refugees. A million of them, from a dozen countries, were wasting away in refugee camps or, like the countless lost Somalis, had become urban nomads, wandering in the capital, many without documents and not allowed to accept paid employment, a Catch-22 situation that showed no sign of ending. I arranged a meeting with the other contact Laura Buffoni had given me, Yusuf Mohamed Warsame, a refugee who had been in Ethiopia for more than twenty years. His father, Somali ambassador to Ethiopia prior to the fall of Siad Barre, had died in penurious exile from his homeland. Since then, Yusuf had become an indispensable leader and facilitator in the Somali refugee community, as well as working on behalf of refugees from eighteen other nationalities. He and his extended family had been approved for resettlement in Canada two years earlier, he told me over the phone, but somehow their case had fallen into a bureaucratic crevasse. Every ten months, they were obliged to pay for new medical examinations, which made it difficult to sustain their faith in the goodwill of what they hoped would be their adoptive country. As our conversation drew to a close, I arranged to meet Yusuf two days hence.

It wouldn't be the first time I'd met Somalis on my journey. Shortly before my visit to Gulu the previous November, I'd made a hectic return trip from Kampala to Entebbe. Dave Copeman of Amnesty International was about to leave for the Oxfam conference at the Imperial Beach Hotel when I arrived at his office. He invited me to join him. Dave was an Australian, good-natured and typically well travelled. As we were chauffeured in the organization's air-conditioned Land Rover through the heat and traffic congestion of the Jinja Road, he told me he'd once spent five months as a ski bum at an Invermere mountain resort near the B.C.-Alberta border in Canada. He was a storehouse of information, recommending a book by Mark Bradbury called *Becoming Somaliland.*

"Somalis, like Ugandans—and people everywhere, for that matter—want five things: food, water, health, education and security," Dave said as we turned into the exquisite gardens of the hotel, "most in short supply, or too expensive."

By the time we reached our destination, the Oxfam conference had ended and lunch was being served. Nonetheless, I met half a dozen Somalis working in Kenya, Somaliland and Somalia proper. They'd been talking amongst themselves about vested interests in the country's potential oil and gas; also, there was a suggestion that Ethiopia preferred a destabilized Somalia to ensure against renewed attacks, such as the one in 1977 when the forces of Siad Barre had tried to reclaim the rich grazing lands of Ogaden. The workers I met had no enthusiasm for another radical Islamic regime in the region. According to one man at my table, Ethiopian spies had infiltrated even the ranks of the Islamic Courts Union, the major contender for power in Somalia.

"Al Shabaab is the problem, not the Islamic Courts Union, which is actually quite moderate," another insisted. "And Ethiopia is giving arms to Al Shabaab."

According to Dave Copeman, local NGOs in Somalia tended to be clan-specific, their principal task the repair of damaged infrastructure. These organizations did good work within a limited sphere, were generally honest and often more efficient in their use of funds than more broadly based NGOs. However, Dave advised, you needed to be aware of the limitations, the fact that the organizations were, in the first instance, answerable to their own people. Another of the aid workers at the table told us about three public executions that had been carried out by police in Baidoa, capital of the Bay area in south-central Somalia. One person had been accused of murder with a pistol, another with a bomb. No mention was made of the victims. The accused were taken to a public place and shot. The local mullah, also implicated, had been dragged from his bed, held in jail, then taken to a remote spot and shot, but not before he had struggled with his abductors in the car, wounding two police officers in the process. The man offering this story also claimed to know the approximate whereabouts of various hostages being held by Islamic fundamentalists in Somalia, including the Canadian journalist Amanda Lindhout and Australian Nigel Brennan, for whose release a ransom of $2.5 million was being demanded.

Dave doggedly questioned this man about details, making sure he had the exact names, times and places. I was glad that Amnesty was on the case. I headed back to the NGO's office with Dave, suggesting I'd make my own way from there. As the Land Rover rounded the final corner, he shouted to one of the *boda-boda* drivers across the road.

"Hey, man! Follow us and you can take the *mzee* downtown." This Kiswahili word, Dave told me, was a term of respect, meaning "the old one."

I looked forward to my meeting with Yusuf, but I had a busy day ahead of me. First, I had to change my room at the National Hotel, where I'd had my fill of electronic evangelical harassment, then deliver my lecture at the University of Addis Ababa.

A threat to change hotels got me a replacement room far from the auditory torture of Kidus Istephanos. Even the view had improved. I could now look northwest over one of the few remaining chunks of parkland. The room was also two floors lower, enabling me to use the stairs. That the cupboard doors did not close and mosquito screens in the bathroom did not fit the window did not trouble me. Any enterprising bloodsuckers and malaria transporters that survived the pollution, heat and altitude of Addis would be either too tired to bother with me or entitled to a good drink.

I crammed some pages of notes and quotations into a folder, bolted down the stairs to the front door of the hotel and waved to Johann, whose cab was stationed across the street. Taxis in Addis were a health hazard. The fleet of ancient Ladas and Renaults, survivors of wars, regimes and multiple owners, was not exactly roadworthy. None had seat belts. Carbon-monoxide fumes drifted in from the ruptured or missing exhaust systems. Most vehicles had 90 degrees of slack in the steering column, so they did not respond well to suggestion. All this made driving in a city with too many cars, too few traffic lights and no apparent rules of the road a challenge. Arriving at an intersection, cars would wait for someone brave or crazy enough to venture into the cross-traffic, then edge out in his wake in a staggered column, like the wing of an advancing rugby team, until they had taken over and established a solid line of through-traffic. Then the whole process began again from another direction.

Having survived several close calls, I had settled on Johann and his taxi as the car-driver combo most likely to get me there alive. His Renault was old but well maintained—even the steering—and had a crucifix mounted on the dash. If I rolled my eyes when he quoted an outrageous price, Johann would laugh and suggest an alternative, just low enough to give me the illusion I was getting a deal. That day we headed north up the divided boulevard past the presidential palace, the Hilton, various government offices, the museum, the zoo and the peripheries of the university, whose buildings spread a mile along the main road. I'd already taken this route several times, by taxi and by bus, meeting Bernahu and attending a book fair with a young journalist named Arefaynie Fantahun, who'd generously walked me through the markets of the city's old Italian section and shown me some out-of-the-way historic buildings.

Bernahu was waiting for me at the university gate. I was nervous, so I appreciated his calming presence. Even after a lifetime of teaching and public appearances, I was always convinced I'd be struck dumb, unable to make a sound. Worse yet, I'd be exposed as a fraud. What did I know about Africa, about politics? The small lecture hall had begun to fill up when we arrived, an audience of both faculty and students. As I put my small backpack on the desk and looked around—there were no narcs or ninjas in evidence—I could feel the telltale signs of what I'd come to call teacher's bowels. I asked Bernahu for the key and fled to the nearest washroom. Sitting there, pants down, brow furrowed, I pondered my situation. Straightening the one tie I'd brought along for the occasion, cursing my inappropriate cargo pants, and patting down my uncontrollable white hair, I made myself visible again to the packed room.

I started by explaining the title of my talk, "Killing the General with Words: Poetry as a Subversive Activity." The "general" in question was Pinochet, I said, and the reign of terror he had inflicted on people in Chile: torture, disappearances, the dismantling of social programs. What should writers do in the face of such terror? Was it fine to ignore events like these, as long as you proceeded with utmost respect for the language? Did content matter and, if so, in what way? Camus had argued that, at a certain point, the writer must put down his or her pen and go to the barricades. Bob Dylan, by contrast, had refused to be labelled as a protest singer. I wanted, I told the audience, to challenge the notion that poetry makes nothing happen, to suggest that some of the best writing is subversive, that it gets under the skin, exposes lies and can change attitudes, maybe even the course of events, release healing potions into the body politic. I mentioned W.H. Auden, who insisted that "the mere making of a work of art is a political act," because it reminds the management that we are human beings, not just workers, and because it refuses amnesia, bearing witness to, and making a record of, our history. And I ended with a quotation from American poet James Scully: "Political poetry is not a contradiction in terms, but *an instructive redundancy*."

The lecture was over, followed by polite applause. A number of hands shot up. Did I have examples of societies changed by poetry? I said the Greeks must have been made more civil by their poets and playwrights, but I could offer no proof other than the emergence of the idea of democracy. I thought Pablo Neruda had borne witness and proven to be a moderating influence in Chile; Ernesto Cardenal had done the same in Nicaragua. Another hand was raised at the back of the room, an elderly gentleman with a bald head. "What about the misuse of rhetoric, or art in general, to distort the truth? As a tool in the hands of murderers and thieves, is art not reduced to the level of propaganda?" This was a difficult question, I acknowledged. The Nazis had certainly made good use of film to advance their goals. Perhaps poetry was like a machete: in the hands of the right person, it could be used to chop wood, build houses. In the wrong hands, well, the results had been obvious in Rwanda. I was in over my head, and too tired to think clearly, so I offered a parting concession. Yeats, I said, liked to distinguish between poetry and rhetoric. He claimed to use rhetoric for his argument with the world and poetry for his argument with himself. I didn't entirely agree with Yeats. My view, I said, is that all poetry is, at some level, political, even so-called confessional poetry, which seeks to manipulate us to respond to and accept the authenticity of a series of events and emotions constructed out of words.

I expressed my gratitude to the audience for their engagement, thanked Bernahu, then slipped away early. I thought I'd have a light meal and a beer, and watch an hour or two of television, but the electricity was down again. Other than a few candles in the lobby and the stairwell, the National Hotel had drowned in darkness. Fitting, I thought. I placed my small flashlight on the table in readiness and curled up in bed, consoled by the prospect of my first Ethiopian sleep-in.

This was not to be. At precisely 6:30 AM, I sat bolt upright in bed to the pulsing rhythm of drums and the high-pitched sound of chanting, as if all the tribes of Africa had congregated beneath my window. I put on my trousers and stepped onto the balcony to have a look. No one was visible in the half-light, but a dozen cars were jammed into the tiny parking space below. From the hotel's ground-floor extension, rays of artificial light anticipated the morning along with waves of murderous sound. The pumped-up bodies of a Pilates class shouted and grunted in unison. I went back to bed with my pants on, recalling my favourite line from Kingsley Amis's novel *Lucky Jim*: "Not for him the slow, gracious wandering from the halls of sleep, but a summary, forcible ejection." I had, it seemed, merely exchanged Orthodox cacophony for the orgasmic cries of gymnastic fundamentalism. I was not amused. I had not expected to encounter devotees of Joseph Pilates in Addis Ababa, but I realized, if I wanted to save my day, I should take his most basic advice: "Even if you follow no other instructions, learn to breathe correctly."

On our way to the Somali refugee centre, Yusuf Warsame and I were intercepted by a small boy of eleven or so who called himself Hossein. Although Yusuf did not know this kid with his bare feet, T-shirt and tattered pants, Hossein had no intention of letting us go or being told to wait. He wanted us to know his story and belted it out in high gear, with Yusuf translating. He had fled the violence in the Toufique district of Mogadishu after his father, a shopkeeper, had been robbed and killed by Islamic Courts militias, along with Hossein's brother. Hossein had made the journey to Addis Ababa by himself, without food and water for days at a time. He told us he had sewn some money into his underwear before he left Mogadishu; when that was gone, he worked as a grease boy on the lorries.

"I ate leaves from the trees. Only when I met people like you," he said, turning his intense gaze from Yusuf to me, "did I have my first drink of clean water. I saw a woman killed with stones. They accused her of adultery. They killed lots of innocent people and captured boys like me to take for soldiers." His rapid-fire account sounded tailored until he realized he was not going be brushed off and could relax.

Hossein had been in grade three in Somalia. Now he was working the streets in Addis, begging. He had collected enough to buy shoe polish, brushes and a couple of rags, but the Ethiopian shoeshine boys had beaten him up and stolen his supplies.

Yusuf translated what Hossein said, then laughed out loud. "This kid is smart. He knows everything." We did not realize just how smart until I asked what Hossein hoped for his future.

Without a hint of irony, he declared: "I want to be the king of Sweden." He launched into a history of those who had risen to power from modest origins, mentioning Barack Obama and Adolf Hitler in the same breath. "Hitler used his own blood to make pictures when he could not afford to buy paint." Geography was not Hossein's strong suit. If Sweden was out of the question, it appeared that any country would do. "Even Canada," he said, "especially if I can walk there."

I asked if there was any hope for Somalia.

"There's no hope to change things now. But maybe, if I get education, I'll come back and help the people." I wondered if he'd thought of going to Somaliland, which was reported to be a safer place. He said he'd been beaten up there because he was from Mogadishu. "I heard once we were one country, but I can't imagine that."

How did he feel about the loss of his family, I inquired. "I can't be crying all the time," he said angrily. "I have to look after myself." Then he admitted that he did not know whether the militias who killed his parents had been Somalis or Ethiopians.

Saida, Halima, Faduma, Agut, Sahro, Hoda. Yusuf had arranged some appointments and a room for me the next day at the Jesuit Refugee Centre, where the space was more congenial for interviews and interruptions were less likely. The line of Somali refugees formed outside the small office, and others, sensing an opportunity, had joined the queue. Halima, age twenty-two, was pregnant out of wedlock, she told me. The aunt who had condoned Halima's forced relationship with a rapist in Mogadishu because of the protection it offered the family now disowned her niece for sleeping with someone she loved. Halima longed for her mother, whose whereabouts she did not know, and realizing she could be stoned for her pregnancy she had contemplated suicide. In desperation, she had gone to the offices of UNHCR and told them: "I will give you the baby." Yusuf was hoping to find her lodgings outside the Somali community, where she could deliver her baby in safety. The other option, not even remote, was that she would win the resettlement lottery and manage to emigrate.

Faduma, an illiterate member of the despised Bantu clan, the untouchables of Somalia, had the gift of song. Her talent was celebrated until the arrival of the Islamic Courts. When she sang on request for a wedding in Mogadishu, she was jailed and raped for three months by her devout Muslim captors. "I was crying. They cut my mouth and raped me beside the dead bodies of my father and mother. They used a bayonet to open me up." At eight months pregnant, she miscarried because of another severe beating. "I reject my country, my roots. I will never go back."

Saida, thirty-five and wearing a beautiful yellow hijab, had lost her father in a tribal clash. Marriage was her only option, but her husband, who owned a small business, had been murdered in front of her. Her life went steadily downhill: a second marriage to a bully, from whom the eldest children ran away; a refugee camp where she lived in fear of this man; a daughter with a chronic heart condition who could get treatment only in Addis, while the other children remained in the IDP camp; education for this child not possible because of the need to commute back and forth from city to camp. Though depressed, Saida was a fighter; that much was clear. She refused to allow her girls to be circumcised, she said, so she was doubly under threat.

"What do I want? Somali women are just chattel, clan property. We have no rights. We are always at risk. If my husband died, I could be taken by the meanest and ugliest fifth cousin. I am tired of thinking, tired of having nightmares about my last husband, tired of wondering where I will end up or what will happen to my children. An end to all that is what I want."

Becoming a refugee is most often a life-or-death matter. The decision to abandon home and belongings for the perils of the road, to place your life and that of your children in the hands of strangers, to surrender your fate to the arbitrary standards of bureaucrats, to become, at once, powerless, an object of suspicion, a creature in limbo, a scarcely noticeable statistic never comes easily. As I looked into the eyes of each refugee Yusuf and Mulugeta brought to me, I thought about the arbitrariness of borders and about the small circles of support we wrap ourselves in: family, friends, work or a profession, ethnicity, a neighbourhood, a country, the circles diminishing in intensity as they move outwards, like the ripples in a pond. Imagine each of these circles shattered. Your hearing is impaired from a blow to the head by a rifle butt; or there's a dull pain inside from the damage of multiple rapes. You've been falsely accused; the mafia that feed on the misfortune of refugees are encouraging you to flee yet again, for a price; even the old *faranji* listening to your story is not sure you are telling the truth about the bishop in the Kivus who hid weapons and ammunition for the rebels under the medical supplies you were delivering on behalf of Caritas Internationalis. Although you know a small boat sank recently crossing the Mediterranean and seventeen graduates from the University of Hargeisa were drowned trying to make it to Europe, you're willing to try, to put your trust in Allah, though desert crossings to Khartoum and Libya hold the very real possibility of robbery, rape, slavery or death. No, you say when asked, I don't know how to swim, but, *inshallah*, maybe I'll be lucky.

When my own mother died, I felt I had been orphaned. My father lived thousands of miles away, married to another woman. There was talk of family friends adopting me, but my father showed up soon after with the intention of taking my brother and me to live with him. I sat at the piano, having had only twelve lessons, but desperate to play well so my father would not change his mind. At the time, I felt I was playing for my life. These young people were playing for their lives, too, hoping to elicit my sympathy and assistance. The subtext in everything they said was the same as the question that Angélique posed to Sam in Kigali: "What does Gary want, does he plan to help me?" Sitting there, recording their pain, I felt the paucity of what I had to offer. These feelings were intensified by my own guilt as a father, of having walked away from the marriage when my first daughter, Jenny, was one year old, of having missed the surprises, the responsibilities, the intimate arc of her maturing. It was as if the selfishness and ruthlessness I demonstrated in trying to make myself acceptable and successful as an adult had been at the expense of this beautiful child who, little thanks to me, is now a gifted and successful woman.

I'd interviewed refugees before: Vietnamese refugees in Hong Kong, boat people who had risked everything on the high seas to reach camps where they would languish, waiting for resettlement; two Palestinian brothers, both blind from vitamin deficiency and intolerable conditions, including open sewers, at Deheishe refugee camp in Bethlehem, the reputed birthplace of Christ; and Afghan refugees, packed into plastic tents in a field in the Northwest Frontier Province of Pakistan, gaunt, haggard faces, but illuminated from within by faith in Allah and a not-yet-extinguished spark of hope. An entire lifetime would not be enough to rescue them all, to provide the care and the refuge they deserve. How do you decide which one or two you will help? The one whose story is the most horrific, or the most convincing? The one likely to require the least help in adjusting? Go ahead, admit it: you're not perfect, you can't help it if you find certain faces more familiar, more appealing, more like you.

And what of the refugee, if the final destination is achieved? Make no mistake, resettlement is not going to be easy. Your engineering degree may earn you a job renting bicycles; your nursing credentials, a broom. Your clan privileges will count for nothing; your brood of precious children will be viewed as a liability. You needn't have read Robbie Burns's poem "The Cotter's Saturday Night" to know fortune is fickle, that "princes and lords are but the breath of kings." Ethiopia, once a land of kings, had become a nation of politicians and paupers.

I was headed for Somaliland, but before I left I shared a few of these thoughts with Elias Cheboud, a builder of bridges between Africa and the West who taught sporadically and worked part-time for the University of Peace in Addis. Elias had earned a doctorate in Canada and been employed as a visiting professor at the University of Victoria. Although he had arrived in Canada as a qualified medical doctor, he was patronized and treated as a pariah or an imbecile by some of his professors.

"In your country, there's a natural assumption of inferiority when it comes to Africans," he said. Children had taunted him in the streets; an old woman had attacked him with her umbrella, calling him a monkey. His own children, from a mixed marriage with a white woman, cursed him for the bad luck of their colour. In Edmonton, Elias had been dragged from his car, beaten and hospitalized with a damaged sternum. The authorities claimed to know the perpetrators, but refused to prosecute. Like Wolde Selassie, he had nothing good to say about aid organizations and their practice of paying local staff—no matter how experienced or well educated—as little as 5 to 10 percent of what they paid expatriates. "I needed to learn these lessons," Elias explained. "I came back here to work for Ethiopia. The income is small, but the satisfaction is enormous."

From my third-floor balcony at the National Hotel, I watched the last rays of sunlight gild the tops of distant buildings. Near at hand, everything was in shadow. All week, I'd felt swamped by a tsunami of need; the stories I'd listened to were red-hot branding irons searing my conscience. Yet when I was confronted with the belief, the gratitude and, yes, the hope in some of those faces, a strange peace had settled on me. I'd come away laden not with guilt, but with the sense of having been blessed and challenged. Maybe I'd learned something, too: that there are moments when it's more important to tend the wounded than to report the casualties.

In the ravine that backed onto my hotel and housed the slums, amaryllis, flowering hibiscus and other exotic species had surrendered their brilliant colours to the advancing night, but continued to release their intoxicating fragrance, liberally, almost recklessly, to the desperate and the privileged alike.

seven *Airborne Particles*

NO OTHER African people have provoked such extreme responses from visitors and observers as the Somalis. Anna Simmons, in *Networks of Dissolution: Somalia Undone*, is typical: "Somalis lie, cheat, and are quick to anger. They are proud, vain, and think highly of themselves. At the same time, they will act courageously, faithfully, and are capable of enduring great hardships as well as intolerable pain." Although this list of attributes might be applied equally to most nations, Somalis often find themselves singled out for such generalizations. The stereotypes applied to them are no less misleading than the oversimplifications and misrepresentations of their domestic politics. As historian Steve Blankenship concluded in 2002, "Somalia stays stranded in the peripheral vision of the West as another indistinguishable African country whose alien culture and incomprehensible politics make it less a nation to be reckoned with than a spectacle to be consumed, then forgotten."

But Somalia has not been forgotten. Battles for control of the Horn of Africa are ongoing. Old enemies, Ethiopia and Eritrea, are deeply involved in Somali politics, the Eritreans supporting and arming the Islamic Courts Union, and the Ethiopians, with U.S. support, serving as an occupying force in support of the Traditional Federal

Government (TFG). The radical group Al Shabaab, meaning "the youth" (clearly modelled on the Taliban, which derives from the word *talib*, meaning student), controlled most of southern Somalia until December 2006, when the U.S. effectively broke the group's hold on power. But the intervention could not defeat the insurgency. By the time of my visit to the region, bloodshed and mayhem had destroyed all semblance of a functioning government. The TFG had fled to Nairobi, there to undergo a falling-out between president and prime minister and virtual paralysis. Only 3,400 troops of the African Union's peacekeeping force, AMISOM, and a small contingent of Ethiopian troops remained in Mogadishu to confront increasing attacks by a variety of militant groups.

On my way to Addis Ababa, I'd had a stopover in Kigali, and while there I'd spoken with Harald Hinkel, recommended to me by several people for his general knowledge of Africa and his work experience in Somalia. Harald, a trained herpetologist, had been doing snake research in the national parks in DRC until the turf war between rival militias became too intense; then he went to work for the World Bank, demobilizing militants. He looked as if he had stepped off a movie set, one of those handsome and solid, muscular types who appeared to have absorbed all the concentrated maleness of the species, and a more than usual chunk of the intelligence. I could feel myself shrink in his presence. Harald also had a sense of humour. He told me about a surprise attack on Gisenye by Mai-Mai militia who had crossed the border completely naked except for the AK-47s they carried. They took the town briefly, but were eventually driven back over the border. Asked by a reporter how a small gang of "crazies" could have achieved such a victory, a young Rwandan private shrugged his shoulders. "I couldn't shoot at a naked man!"

Harald had been in Africa at the start of the Rwandan genocide, but was shipped back to Germany with an advanced case of malaria. His wife, a Tutsi, took refuge in the famous Hôtel des Mille Collines in Kigali. All the room phones had been removed, he told me, but she managed to place a call to Germany through a phone/fax machine in the office that had somehow been overlooked. "It's bad here, get me out," were the last words she uttered before the line went dead. Miraculously, she survived the slaughter. Harald now divided his time between Kigali, where he lived with his wife and children, and the two provincial capitals of Goma and Bukavu. The demobilization work was challenging, especially with 106 groups of "nasties," the worst of which was a militia known as the Rastas, comprising FDLR and Mai-Mai. I restrained myself from remarking that human behaviour can be lower than that of snakes. A similar offhand remark earlier had earned the prompt reprimand: "That's not the point!"

Harald acknowledged the complicity of mining companies in the conflict and explained how they avoid paying taxes or royalties for years by prolonging the so-called exploration stage. These samples, of course, go out in huge quantities. I likened this to the claims of the Japanese that they are whaling for scientific purposes.

From DR Congo, our conversation shifted to the other coast. Harald had spent three and a half years in Somalia, where he was shot in the throat by a member of one of the clan militias. Although he'd almost died, he had great admiration for the Somalis as warriors. They were fierce fighters, he said, but logistically challenged, impossible to command. Somali troops had struck fear in the Japanese enemy and driven their own British officers mad in Burma during World War II.

"They're the finest and most generous people in Africa," Harald said, "but you can't believe a thing they say." He ascribed this paradox to conflicting loyalties—family, tribe, clan and nation—in descending order of importance. Then he recounted a long exchange in a grassroots court between himself and his assailant, after the man who shot him had clearly been caught dissembling.

"Why did you lie to me?" Harald inquired.

"If I told you the truth and you did not believe me, what would I have as a fallback position? I saved the truth for later."

I had Harald's and Steve Blankenship's words and the issue of reliability very much in mind as I waited to board the plane to Hargeisa from Djibouti City. To get to Somaliland, I'd had to spend two nights in neighbouring Djibouti, once called French Somaliland but officially known after independence, from 1967 to 1977, as Territoire français des Afars et des Issas. In addition to being an active international port, Djibouti was still very French, not only in terms of its high prices, but also in the abundance of restaurants and good food, including cartloads of fresh baguettes exuding their aroma on street corners. The only English television channel was Al Jazeera. Snippets of Russian and Chinese were as likely to be spoken on the street as English, though a hole-in-the-wall hair salon and cosmetic shop across from my hotel carried the inauspicious name She Stylish. I spent hours walking around the city enjoying the sights, only to be accosted by a young zealot offended by my camera.

He touched my arm. "We don't want you to take pictures."

"We?" I knew what he meant, but did not appreciate his rudeness and the plural pronoun.

"You must not take pictures. This is a Muslim country."

"The Qur'an does not prohibit images of landscape and buildings, only of people."

He hesitated for a few seconds, then announced: "What you are doing is not permitted."

By whom? I asked in French. "Est-ce que vous êtes la police?" He became very angry. I understood his point of view, I told him, but he insisted on following me down the street a short distance, giving up only when I entered a shop.

At the airport, I was startled to discover my flight was not listed on the departure screen. Several other passengers, sharing food or quietly masticating mouthfuls of *khat*, assured me they were on the same flight to Hargeisa. Only when I found a room the size of a broom closet with the Daallo Airlines sign on the door did I begin to relax. The solitary clerk was reading a newspaper, but gave an affirmative nod when I inquired about the flight.

I regretted not having had the time to explore Djibouti's lunar landscape, where *Planet of the Apes* had been filmed. However, I was able to see it briefly from the air before the windows of the old four-prop

Antonov fogged up. The Soviet relic, painted white with a bruised blue nose, had been overbooked, so there was a mad rush for the open seating, after which a dozen or more desperate unfortunates paced the threadbare carpet in the centre aisle before being ushered off the plane for a two-day wait. A plump Russian mechanic in tattered jeans and a dirty, sweat-soaked T-shirt emerged through a small opening in the fuselage amidships, pulled up a ladder, then disappeared into the cockpit, closing the door behind him. The seat belts were not adjustable, the recliner mechanism was kaput, luggage racks did not exist and the nearest emergency exit was ringed with blue masking tape. Seated a few rows ahead of me, a distinguished-looking gentleman was holding the toilet door closed with one foot. The Antonov lumbered along the runway, became airborne with less than a hundred yards of tarmac to spare and, with labours worthy of Hercules, struggled inch by inch to gain altitude. Putting pieces of the puzzle together, I realized that the Daallo Airlines flight had not been listed on the departure board because the plane did not meet civil aviation safety standards. It was being flown illegally, and uninsured.

Boobe Yusuf Duale was there to meet me on the tarmac, his assistant whisking away my passport and the necessary visa fee. A small crowd had gathered and were watching the remaining passengers descend the stairs from the Antonov.

"Welcome to Hargeisa, Gary! We'll get things sorted out for you in no time at all."

I'd contacted Boobe through the good offices of Martin Orwin, who teaches in the School of Oriental and African Studies at the University of London and is a leading expert on Somali poetry. Martin, I knew, had arrived in Hargeisa a few days earlier. I did not kiss the ground, but happily baked in the intense sunlight. Boobe excused himself briefly and went to greet another passenger arriving on the same flight. It was the gentleman who had kept his foot propped against the toilet door. A cheer went up from the crowd.

"It's Hadraawi," Boobe announced a few minutes later. "You'll meet him in a couple of days." Just as well, I thought, enough time to get my wits about me. A poet with the status of a rock star, and one of my reasons for being in Hargeisa, Hadraawi was a national treasure. He stood on the tarmac, smiling and shaking hands, dressed in a blue shirt, a patterned red bandana around his neck, an embroidered Muslim cap and a *macawis*, the traditional Somali wraparound skirt. He was already besieged by well-wishers. Hadraawi had been a strong critic of the Siad Barre regime, which imprisoned him between 1973 and 1978 for not toeing the party line. Free again, he served five years as arts director of the Academy of Science, Arts and Literature in Mogadishu before quitting to join the Somali national resistance movement based in Ethiopia.

Although a sovereign state briefly after it achieved independence from Britain in 1960, Somaliland had relinquished its new-found freedom to join the greater Somali republic, a regrettable decision since the larger nation soon descended into dictatorship, chaos and war, including an abortive attempt to win back the disputed Ogaden grazing lands from Ethiopia. In 1991, after fierce resistance and the eventual overthrow of Siad Barre, Somaliland declared itself independent once again, though seventeen years later it still remained unrecognized by the international community. The region had been brutally attacked by Somali government forces, with bombing raids that levelled the capital, Hargeisa, and claimed fifty thousand lives. As we drove through the downtown area, Boobe pointed out the MiG jet fighter mounted on a pedestal near an important intersection. Compared with the size of today's arsenal, it looked like a toy.

"That's one of the planes that bombed us," Boobe announced with pride. I'd discover no shortage of reminders of war in the city, men with missing arms and legs who had lost their means of making a living, along with everything else.

"In 1991, the population of Hargeisa was only 30,000," Boobe explained, just missing a wheelbarrow heaped with bananas. "Now it's over 700,000." These numbers, and the lack of international recognition and support, explained the dust, potholes, paucity of paved roads, fledgling public services and the fleet of donkey carts, each laden with a discarded fuel drum full of water to be sold in small amounts to the poor. As we passed one of these contraptions, which had stopped to fill a woman's blue plastic container, the driver put the hose into his mouth to start the suctioning process.

The car slowed to a crawl, as pedestrians, merchants, money-changers and idlers moved aside to permit our passage. Boobe explained that he had studied geology in Italy, but had been too busy with anti-American protests and with cheering on Stokely Carmichael, Angela Davis, Eldridge Cleaver and the Black Panthers to finish his degree. This was followed by an intensive political science course in Moscow.

"I loved the fairs in Milano," Bobe said as we pulled up in front of the Oriental Hotel. "They used to ask me where an African learned to speak such good Italian. I told them it was from eating so much spaghetti." After I'd checked in, we ordered lunch in the lounge.

Boobe had worked in the ideological unit of the Siad Barre government, in the literature section, he told me, but quickly realized that northern Somalia would always be neglected. "My bosses kept asking me why I write so much about Hargeisa," he laughed. He defected during an official trip to Geneva, then took refuge in Ethiopia, where he used his old contacts to raise support for the rebellion. He was appointed as a central committee member of the Western Somali Liberation Front, fighting against both Siad Barre and the government of Ethiopia in the Harshin region, just inside the Ethiopian border.

"We started with one truck. The leader sat in the cab; I rode on top." Boobe had worked mainly in propaganda: radio, film footage, plays, music, agit-prop stuff needed to sensitize people to the issues of resistance. He still kept his old gun at home as a souvenir. "I grew up less than a kilometre from this hotel. By 1991, when Somaliland declared its independence, this was a ghost town, and mostly rubble."

Boobe invited me to visit him at the office of the Academy for Peace and Development the following morning. When he stood up to leave, I noticed he had not touched the plate of spaghetti he'd ordered.

In the early morning, I spent two hours prowling the open markets surrounding the Oriental Hotel, full of fruit and vegetable stalls, charcoal burners and hanging racks of colourful fabrics through which the sunlight filtered, creating a giddy, kaleidoscopic effect. Unlike the man in Djibouti, Somalilanders were not offended by my camera; instead, many asked to have their picture taken and would drag their friends into the frame. Several women did cover their faces and turn away, or wag a finger to discourage my prying mechanical eye, but this was often accompanied by an amused expression. As I paused to adjust the camera, I felt a nudge on my ass. A light brown, cud-chewing, window-shopping cow that had shadowed me for a hundred yards, sniffing tables of fruit, racks of running shoes and brassieres, was fed up with my slow pace and anxious to get past.

Taking my cue from this impatient bovine, I hailed a cab for the short trip to the office of Maria Vargas at the Danish Refugee Council, where I hoped to get a perspective on the human rights situation in Somaliland. Maria was neither shy nor cautious on the subject. In fact, she launched right in with booster rockets firing.

"Fighting for human rights here," she told me, "is like trying to push porridge up a slope. In Somaliland, the victim of a rape is often forced to marry the rapist, and revenge is common; blood feuds go on for years over something as inconsequential as a stray goat wandering into Granny's garden. No apologies or punishment for a murder, just compensation by the relatives or clan: blood money, usually in the form of camels or other livestock. Never mind Western criminal law, this is an ancient land, a moderate Muslim country where customary law supersedes even shari'a."

You could not expect recognition of human rights, Maria explained, when major and minor clans were segregated, when minority ethnic groups, such as the Bantu and Gabooye, were persecuted and seldom had access to education. And women's rights—forget it. Female genital mutilation was still practised. "Women have no marital property rights: they are mere chattel. Employment opportunities for women are few; their political clout is minimal." It was a far cry from Rwanda, where women politicians were in the majority; Somaliland had only two women MPs out of twenty-two, and even those two seats were viewed by some as an outrage. Yes, new political parties existed, Maria told me, but no new thinking, no evolution of values. Most of the ambitious returnees were misfits, interested in power but with no commitment to building infrastructure or creating civil society. Impunity was the name of the game. As an oral culture, there's plenty of talk but little action, she said. And, for many, no collective memory. To illustrate, Maria told me that SHURANET, the Somaliland Human Rights Network, had been closed down by the current administration because it challenged the government's emergency law, which permitted arrest without warrant. The country staggered along thanks to remittances of a billion dollars a year from the diaspora, which accounted for a greater portion of the GDP than did the livestock trade.

"Children? Don't get me started." Elbows resting on the table, Maria rocked her head in her hands. She cited the case of a nine-year-old boy who spent two years in an adult prison because of a custody battle between his parents. And then there was the problem of how to feed, house, nurse and educate the legion of displaced newcomers, mostly from Somalia. Somaliland called them refugees, but UNHCR and the international community and its NGOs didn't recognize Somaliland as a state and so considered these people to be IDPs, making them ineligible for full assistance. Awareness campaigns were under way, but the task of educating people about their rights had proven enormous, even without the problem of clan loyalties. "We do our best," Maria sighed, "but it's not easy," a phrase I'd heard in every country I'd visited. A respected Oromo leader was recently deported back to Ethiopia, she informed me, because he dared to challenge the Somaliland government's decision to remove refugees from the former State House to a location ten miles outside Hargeisa, where they would be out of sight and out of mind. After we'd made several orbits of the politically troubled earth, Maria returned to the subject of incompetents seeking positions of power, some from Mogadishu, others from abroad, most of them obnoxious and ill-informed. One new minister claimed Somaliland was not bound by international human rights agreements it had not signed, unaware that his government and the previous administration had enshrined those rights in the constitution. On a positive note, her office was funding an anthology of local poems about human rights edited by a young writer named Hassan Madar.

As I was being shuttled by taxi to Boobe's office at the Academy of Peace and Development, the driver hit the brakes abruptly as he approached an intersection. A high-pitched ambulance siren was bearing down on us. A late-model blue pickup fishtailed around the corner, almost clipping our vehicle and sending up a huge plume of dust. "What was that?" I asked, both hands braced on the dash. "Some sort of emergency?"

"You might say so," the driver said. "It's the *khat* wagon, making its daily deliveries."

Given the level of addiction to this mild narcotic weed in Hargeisa, a failure of the major supply vehicle to arrive on time might have provoked a municipal shutdown, even a riot. *Khat*—whose active compound, alkaloid cathinone, acts like an amphetamine—made people both less aggressive and less productive, much to the dismay of wives and mothers, whose *khat*-chewing husbands and sons began masticating the substance in the late morning or early afternoon, rendering them useless and grinning in the shade, teeth and lips green from the leaves. I had learned already to avoid drugged cabbies; if enlarged pupils were not a dead giveaway, the fluorescent green teeth certainly were.

The mission of the Academy of Peace and Development, which was founded on September 15, 1998, sounded simple enough: sustaining peace and promoting development, which included facilitating dialogue on human rights, democracy and good governance. But consensus building, like the promotion of human rights, was anything but simple in Somaliland. Teams from APD travelled around the country talking to youth, police, local administrators, women's groups, clan elders and farmers about a host of issues: a trip to distant Erigavo included efforts to address problems related to drought; in Burao, security and the promotion of peace-building initiatives; everywhere, the social and environmental impact of high winds, soil erosion and temperatures above 113 degrees Fahrenheit. Land ownership ranked high on the agenda not only as a way of settling disputes, but also as a means of encouraging industry and development.

In contrast to the bleak picture Maria had painted, the projects under way at the academy hinted at improvements, however incremental, that reflected the local versus the international approach to aid. APD had celebrated International Women's Day by organizing an event that brought men and women together to find ways of eradicating gender-based violence. This involved some frank talk about contraception, the dangers of female genital mutilation and the pressure women feel to vote for relatives or clan members rather than competent female electoral candidates. I could see from the staff I met and the materials Boobe gave me to read that the academy recognized women for their achievements; APD also involved itself in such civic activities as teaching conflict resolution and negotiation skills and registering and educating voters, using scripted videos produced by the organization's film unit. The academy's activities covered all avenues of human experience, including the arts. An Art for Peace photograph on display in the office showed a small group of girls, each holding up the painting she had done for International Peace Day. I was particularly taken with one that depicted a bird on a branch with four marshmallow clouds overhead and "The Peece" written underneath. The pride, hope and intelligence in those young faces struck me as a promising sign for the future.

A tall, bearded German scholar named Markus V. Höhne was using the organization as a base for his continuing research in the Horn of Africa. Markus had developed considerable expertise on the social and political conditions in Somaliland and the neighbouring province of Puntland. He offered to give me some insights into the administration of justice in these regions, particularly the struggle to reconcile the sometimes conflicting demands of Western, customary and shari'a law. Although the nineteenth-century British explorer and author Richard Burton had declared the Somalis to be a "fierce and turbulent race of republicans," he and subsequent scholars also noted a strongly democratic strain that favoured public consultation and negotiation over revenge. Sultans or sheikhs did not decide the outcome of such negotiations. Most disputes were settled by a *shir*, a meeting of elders, not unlike the Afghan *loya jirga*. Each participant was free to voice his (or in very rare cases, her) opinions, though according to Markus wealth, rhetorical ability and knowledge obviously proved assets. *Diya*, or blood money, he said, was not simply a licence to kill. If your family or clan was obliged to pay for damage you'd done—whether injury, murder or theft—extreme pressure would also be brought to bear on you to change your behaviour or face ostracism.

At the political as well as the domestic level, Somali elders had much to contribute to justice and good governance, Markus explained. Those selected to serve in the House of Elders, Somaliland's equivalent to the Senate or House of Lords, were expected to have a good knowledge of the religion and an intimate grasp of Somali culture and traditions. While there was no guarantee that elders would be more resistant to temptation or clan pressures than any politician or adviser might be, evidence suggested that they were a stabilizing influence on elected officials. Debate continued, Markus said, especially amongst opposition parties, who often saw the House of Elders as a rubber stamp for the status quo. But his research indicated that, while suspicions of self-interest and irresponsibility still clung to Somaliland's politicians and to the justice system, and while clan-based loyalties made consensus difficult, the elders were generally regarded as working for the public good.

IN ADDITION TO pushy cows in the marketplace, Hargeisa had on offer a strange assortment of non-human creatures that appeared and disappeared along the streets. Goats were ubiquitous, more common than dogs, and had assumed the role of household pets, especially favoured by old women. No one wanted to eat them, especially as they spent the day scrounging through garbage heaps and consuming Styrofoam containers. When I emerged from the Internet café I'd found, a cavernous affair in a basement off the main street, a small baboon brushed my leg. Without a sidelong glance, it raised one hand over its head, like a basketball player delivering a hook shot, and deftly snatched a banana from a vendor's table. The woman did not react as the baboon sat down less than ten feet from the scene of the crime to consume its loot, deftly and ceremoniously peeling the skin back.

And camels. More than once, a stray camel appeared from between buildings to cross the road in front of the taxi or minivan carrying me through the city. I loved these creatures as much for their ungainly bodies and adaptive skills as for the bad rap they'd received abroad. On my third day in the city, I'd attended an event called the Festival of the Camel at the Hotel Mansour, an extravaganza of poetry readings meant to celebrate the strange beasts that have been the butt of so much humour in the West, where the camel is often dismissed as a horse designed by committee and cursed with foul breath. Here, camels were loved, not only for their amazing physiognomy, which allows them to store water and withstand extremes of temperature, but also for their milk, their meat and their importance in farming, transportation and the military. Roman and Persian armies used camels as mounts and freight animals; so, too, did the United States Army in California in the nineteenth century. The Somaliland Camel Corps fought in battles against Sayyid Mohammed Abdullah Hassan, the so-called Mad Mullah, from 1900 to 1920, then again in World War II against the Italians. It has been said, at various times, that there are more camels than people in Somalia; they once numbered 6.5 million, more than 50 percent of all the camels in Africa.

A recent UN humanitarian report quotes Yassin Mohamed Ganni, a senior elder, as saying that the drought has been so severe in parts of Somalia, with no rain for two years, that camels are dying in some villages. "Camels are the last animals to die and, once they start dying," he said, "it is only a matter of time before people start dying."

I'd read that a Somali's earliest inklings about sex were likely to have come from observing the dramatic reproductive habits of camels. According to L. Skidmore, quoted in *Recent Advances in Camelid Reproduction*, "Sexual behaviour [of male camels] is also characterized by exteriorisation of the soft palate . . . The protrusion of the soft palate occurs all day long at intervals of 15–30 minutes and is accompanied by loud gurgling and roaring sounds. The protrusions become more frequent with increased excitement such as the presence of other males and females. During copulation the soft palate ejection may be replaced by grinding of the teeth and frothing at the mouth." Constrained from writing directly about either sex or the beloved, the Somali poet frequently waxes rhapsodic about the beauty and virtues of the she-camel. His indirect message is unlikely to be lost.

A young woman in the festival audience had raised her hand to ask why male poets still insist on celebrating the camel though they now live in urban environments where the camel is little more than an occasional nuisance. This question provoked a good deal of laughter and a few rude asides. If there are said to be twenty-seven Inuit words for snow, the Somalis beat that record with at least forty-six words for the single-hump dromedary. I'd seen the following examples posted online by linguist Mark Liberman:

abeer or *ameer* "female camel that has not given birth"
awr "male pack camel"
baarqab "stud camel"
baloolley "she-camel without calf that will or will not give milk depending on her mood"
booli "looted camels"
buub "young unbroken male camel"
dhaan "camel loaded with water vessels"
dhoocil "bull camel; naughty boy/girl"
garruud "old male camels; old people"
gool "fat camel"
guubis or *guumis* "first-born male camel"
gulaal "male camel unable to project the gland in his mouth; person with hesitant or stammering speech"
irmaan "dairy camels"
kareeb "mother camel kept apart from her young"
koron "gelded camel"
labakurusle "two humped camel"

mandhoorey "lead ~ best camel in the herd"
nirig "camel foal"
qawaar "old she-camel"
qoorqab "uncastrated male camel"
rakuub "riding camel (from Arabic)"
xagjir "milk-producing camel that is partially milked
(two teats for human consumption, two for the calf)"

The evening's events had been conducted in the Somali language, so what little I'd grasped came from intuition or the occasional translation by Martin Orwin, a fluent Somali speaker and the principal translator of both Hadraawi and Gaarriye. After four hours of non-stop camel poems, we were still not over the hump, and I was definitely feeling *garruud* and *gulaal*, old and tongue-tied, myself.

LATE THE NEXT afternoon, I shaved and rested up at the Oriental Hotel, watching the shadows lengthen, the end-of-day hustle of the street vendors and money-changers with their huge stacks of nearly worthless paper shillings: 6,700 to one U.S. dollar. Here and there, a water-cart donkey brayed to protest its long hours of service. I could also hear the competing calls to prayer of several mosques in the downtown area. I had a dinner engagement with Boobe and a man called Nine, who seemed to be known by everyone.

"Yes, Nine was my teacher," Boobe told me, as we drove to the restaurant together. "He was very entertaining. And he seemed to know everything." Boobe described being sent from class to Nine's home once to fetch a Thermos of tea Nine's wife had prepared. The house was half a mile away and the day was very hot. Boobe was so thirsty he took one drink from the Thermos, then another, before leaving it on the windowsill of the teacher's office. Nine forgot whom he had sent on the errand, and when he discovered his Thermos half empty he demanded to know the identity of the tea thief. "I never owned up to my crime until this morning," Boobe confessed, "when I talked to him on the phone, forty-eight years after the fact."

Nine was tall and in his seventies. Once we were seated in the outdoor patio of the restaurant, where bamboo and tall shrubs cast shadows on the faces of my two hosts, I could not resist the obvious question about his name.

"I was a new boy in class, only nine years old when the others were twelve or thirteen. The teacher embarrassed me by telling the other boys that the new nine-year-old would beat them all in the exams. As his prediction proved accurate, I was thereafter branded Nine."

"It's a lucky, even magical number in Chinese lore," I said.

"Not so everywhere. It could be mistaken for the 'no' in German."

Nine was a handsome, vigorous man. I reminded him of the whimsical promise he'd made to another former student, decades earlier, to offer special classes in seduction. In good Somali fashion, Nine had approximated the indirect method of discussing sex, quoting the phrase "Qabatinta Neefkay taqaan": "A jackal is sufficiently skilled to know the vantage point from which to tackle the goat."

"Now where did you get that information?"

"I have my spies," I said. Boobe laughed so hard he choked on his food. The conversation shifted to Said Samatar, my source, now a professor of history at Columbia University in New York. During my research, I'd come across his book *Oral Poetry and Somali Nationalism: The Case of Sayyid Mahammad 'Abdille Hasan*, about the so-called Mad Mullah who had led a successful and sustained insurrection against the colonial Brits. I'd been struck not only by Said's scholarship, but also by his wit, particularly a reference he made to the Westerner's inability to come to terms with the Somali pastoral habit of stealing livestock, especially camels. Westerners, he said, do not understand the "fluidity of ownership" in Somali culture.

Nine and Boobe were delighted by this phrase, which challenged the notion of private property. "You taught him well," Boobe said.

Nine flicked away an invisible fly. "I refuse to accept the blame." He was serving me from the three platters of food that had arrived on the table, which included a goat-meat stew, a spicy vermicelli dish with raisins and almonds, and a breadlike pancake called *muuffo*.

Nine licked his lips, then said: "Samatar is incorrigible. It's clearly a case of nature, not nurture."

Said Samatar was born amongst the camels and did not know his exact date of birth. He had moved to the Ogaden at the age of sixteen, where he met his father for the first time. "What is your name?" his father asked. Said told him, but that was not enough. "Which of my wives was your mother?" Said remained for several years in the Ogaden, where his formal education took place. He taught in Mogadishu until he received a private scholarship in 1971 to study in the United States. There, he had explained to me in a letter, he learned to be a welder, while pursuing degrees leading to a doctorate from Northwestern University in 1979. When I wrote to him of my plans to travel to Somaliland, Said had insisted that I look for his old friend Nine.

"Any café or restaurant in Hargeisa will do. Everyone knows him."

Said had sent me an amusing essay he'd written for a Somali wedding in Bethesda, Maryland, in 2005, the gist of which I shared with Nine and Boobe, about the origins of the hijab:

> To my knowledge, there are only two references to veiling in the Koran (XXXIII, 53 cf XXXIII, 32) and both pertain to the exceptional females of the Prophet's (PBUH) exceptional household. No injunction from the sacred scriptures of the Koran enjoins the generality of Muslim women to veil. Veiling as a cultural practice comes to the Muslims from Persia where it was the pre-Islamic habit of pagan Persian upper class ladies to veil in order to distinguish themselves from the "street rabble." Therefore, the custom of veiling became a widespread practice among Muslims only after the incorporation of Persia into the Muslim world a century later. Thus, it turned out to be a matter of poetic justice that where the Muslims succeeded in making a religious conquest of Persia, replacing their antique Zoroastrian belief systems with the tenets of al-Islam, the Persians took their revenge by making a cultural conquest of the Umma, or the Islamic world community of faith, through the universal imposition of the Black Monstrosity.

> And now, to the symbolic misfortune of the unfortunate Somalis, with the collapse of the Somali state and consequent apocalyptic cataclysms that roiled the Somalis, the Saudi Wahhabis with their big money and primitive ways have invaded unguarded Somalia with this malignant custom. Hence, where you used to see Somali women in elegant *dira'is* and the provocative partially-exposed take-a-peek-at-my-left-breast *guntiinos*, you have to settle in today's Somalia for a parodistic parade of weird Black Apparitions or, to borrow a Somali expression, *"Oo sidii Gorayadii u humbaalaynaya."* "Slinking about like ghastly male ostriches!"

Nine was nodding his head. "Yes, we seem to absorb the best and worst of the cultures that invade us," he said.

It was dark now. I could hear a dog barking and the soft hum of exposed wiring. There was no time to share another Samatar witticism, which seemed to me so telling: "I say, God or the devil or whatever capricious agencies that created and run the world, played a cruel practical joke on us Somalis, making us uncommonly attractive creatures and yet withholding from us the balancing gene to get along. Can anybody divine the ways of the curmudgeon called Allah?"

As I listened to my companions talk about old times, laughing and plying me with food, I realized that most of the stereotypes about this beleaguered country and its people had been initiated by the Somalis themselves, thanks to their wit and great capacity for self-mockery. Dinner with Nine and Boobe had come to an end, the excess of food swept away. It was as if an extra portion had been prepared to accommodate our invisible guest, Said Samatar, around whose irrepressible presence so much of the conversation had circled.

AIDAN'S STORY, WIDELY known in the Somali community, was news to me. I'd been having lunch at Hotel Mansour with the staff of HAVOYOCO, the Horn of Africa Voluntary Youth Committee, a local NGO that provided a wide range of vocational training for street children, single mothers and the otherwise disadvantaged. After lunch, I mentioned Aidan's name and his uncle, a sheikh from Dubai who had joined us, phoned his nephew and invited him over. Ten minutes later, Aidan and I were knee-deep in the problems of Somaliland. "Somalis understand violence and death, but not human rights," Aidan said, shaking his head in exasperation as we sat in the lounge. "We need education. We need international experts to counteract the brain drain, and for capacity building, for mentoring and for strengthening civil society. Every sector needs help: women, youth, those with disabilities, HIV/AIDS, minorities, child labour, the environment." He paused, eyebrows raised, and extended his open palms like two parentheses in a gesture of helplessness, and to indicate that the list could go on and on. When I asked what had brought him to this point and what kept him going, his personal narrative came spilling out.

After he'd had two years of medical training, Aidan and his friends decided to do something useful in their war-ravaged community. They set up the Hargeisa Self-Help Group, known locally as UFFO. Initiative and independence of spirit were two things the ruthless regime of Siad Barre would not tolerate, so Aidan and his friends were arrested in November 1981 for volunteering their services to the Hargeisa Group Hospital. They were tortured and placed in solitary confinement; twenty-eight bright young men, who had put their skills to work rebuilding civil society, providing counselling and medical assistance, locked away as enemies of the state, incarcerated in a Dickensian maximum-security prison known as Labaatan Jirow. I'd passed this prison earlier in the day along one of the city's main drags; it was now rebuilt with high white walls and a presumably less gruesome interior, a short distance from the newly opened Obama Restaurant.

Time in solitary confinement was measurable only by the yearly change in diet during Ramadan, Aidan told me. If your physical health did not crack first, your mental health would. To stave off madness and despair, Aidan and his associates had developed a unique form of Morse code, tapping softly on prison walls with their knuckles while listening for the steps of the guards. It took hours at first to construct a sentence in this "wall alphabet," but eventually it became familiar and served as a kind of shorthand. After several years, Aidan was permitted to speak to the warden, who offered him new clothes and the choice of one book from the library. Aidan chose the biggest volume he could find, Tolstoy's *Anna Karenina*, at eight hundred pages. He devoured this novel and shared it, via the difficult and sometimes painful wall alphabet, with a friend in the next cell who had become sick, discouraged and suicidal.

"After two months, he started to recover." Aidan smiled as he replaced his coffee cup on the table in the hotel lobby. "Tolstoy saved our lives, but I can no longer bear to read his novel."

Eight years passed before the men were finally released.

As we exchanged contact information, I mentioned to Aidan the parallels between his story and that of jailed Argentinean writer Jacobo Timerman, author of *Prisoner without a Name, Cell without a Number*, whom Phyllis Webb wrote about in her poem "Prison Report":

> In this prison everything is powered electrically
> for efficiency and pain. But tenderness is also
> a light and a shock.
>
> An eye, a nose, a cheek resting against a steel door
> in the middle of the dark night.
> These are parts of bodies, parts of speech,
> saying,
> I am with you.

Aidan shook his head, then embraced me.

SOMALIA LIES IN a region that has been strategic at least since the opening of the Suez Canal, when Britain, France, Italy and other contenders for the riches of Asia desired bases along their trade routes. The partitioning of the fiercely independent Somalis by colonial powers disrupted clan alliances and the imposition of arbitrary borders resulted in more conflict. Britain's transferring of Somalia's Ogaden grazing lands to Ethiopia further destabilized the region. Like the Afghans, Somalis had never needed or wanted to be governed centrally. Unlike the Afghans, however, who contend with different languages, ethnic mixes and competing versions of the truth, Somalis all spoke the same language and were more or less resistant to radical or disruptive strains of Islam. Cold War politics and the imposition of a unitary state on an illiterate and largely tribal people were destined to fail.

When the dictatorship of Siad Barre collapsed in 1991, warring factions competed to fill the vacuum, with no shortage of arms left over from the Cold War. Thirty thousand deaths were reported from the fighting. Drought, then flooding joined the mix, resulting in famine and displacement. Tens of thousands of desperate, malnourished people staggered along the roads to Kenya and neighbouring countries, pushing the weak in wheelbarrows. Images of gaunt, starving women and children began to appear in foreign newspapers. International relief agencies responded, but food distribution was disrupted by rival gangs, many of whom were hoarding supplies, sending truckloads of food to Ethiopia in exchange for money or weapons. Pressure from the world public and the United Nations, along with the fear of Somalia becoming a breeding ground for radical politics, led the United States to intervene in August 1992 with Operation Provide Relief. U.S. Air Force C-130s delivered 48,000 tons of food and medical supplies, much of which was hijacked. A military mission to support the humanitarian effort called Operation Provide Hope was launched.

The debacle that occurred when U.S. forces tried to take out clan leader General Muhammed Farah Aideed, considered a spoiler, is well documented in scholarly articles, news reports and popular books such as *Black Hawk Down*, which was made into a feature film. The Canadian Airborne Regiment, based in Beledweyne, 180 miles northeast of Mogadishu, played a small part in the unfolding tragedy and a rather large part in bringing me to Africa. Canadian soldiers killed one Somali, execution-style, and wounded another outside their compound in Beledweyne. Shortly after, on March 16, 1993, Somali teenager Shidane Arone was lured into the compound, tortured and beaten to death. When the news reached Canada, followed by revelations of the racist hazing videos, trophy photographs of a dying Arone, military cover-up and a commission of inquiry that was stopped in its tracks before officers and top military brass could be implicated in the affair, the Canadian Airborne Regiment was disbanded in disgrace.

As it was too dangerous for me to visit Beledweyne, I was determined to go into the countryside outside Hargeisa, where I thought some of the events of that tumultuous period might become clearer in my mind. With the help of Boobe, I hired the obligatory car and two armed guards and, in the company of Ahmed Keyse, a self-appointed guide I'd met at APD, set out on the three-hour drive to Berbera, a small port town on the Gulf of Aden. Suicide bombers in Hargeisa the previous October had killed twenty people, caused significant damage to property and forced a general tightening of security. While tourism was not yet a serious business in Somaliland, the murder of a foreigner would not help the public image of a stable society desperate for international recognition. Arid conditions were difficult enough here, a few hours' drive from the gulf and the Karkaar Mountains, without adding the threat of military activity. The situation would have been even worse hundreds of miles south in Beledweyne when the Canadian troops arrived in December 1992.

The bleak semi-desert of Somalia, with its intense heat, blasted shrubs, rocky outcrops and packs of scrawny camels, would not be the posting of choice for foreign troops, especially when they were regarded with suspicion, if not open hostility. When news broke of the murder of Shidane Abukar Arone, during a mission with the now sadly ironic title of Operation Deliverance, I had been appalled, but certainly not surprised, given that the culture of war trains men to kill and demonizes the enemy. Armed forces have always attracted more than a few bullies, racists and psychopaths. What troubled me was the response at home—military brass denied the reports, tried to whitewash the murder of Arone and another execution-style "incident," lied to the public, then leaned on the Canadian government, on the grounds of national security, to terminate the inquiry. Equally galling was the readiness with which my fellow citizens and the media got onside, dismissing the atrocity as a case of a few bad apples. Nothing, it seemed, could convince the public that racism was a systemic problem, a nasty aspect of Canadian culture, denied by whites but old news to First Nations and visible minorities from abroad.

In the words of Herschel Hardin, a playwright and professor of political science at Simon Fraser University, "Creating a simulacrum of innocence is only a way colonials have of avoiding their condition." Just as the racism behind the Somalia Affair prompted me to read African history and decide to make this journey, so, too, it had become clear to me during my travels that racism is a key factor at the heart of Africa's problems. The failure of Western countries to intervene in the Rwandan genocide, to take the steps necessary to address ongoing violence and impunity in DR Congo, Zimbabwe and northern Uganda, to require that good social and environmental standards be adhered to by foreign companies doing business in Africa: all these lapses are proof positive that racism is alive and well in the West and that African resources are considered more important than Africans themselves.

Before leaving for Africa, I had talked at length about the Somalia Affair with Toronto photojournalist Jeff Speed, who showed me photographs he had taken during the week he spent in Beledweyne. Speed was the first media person to learn about the murder of Shidane Arone. He was also on hand when Corporal Clayton Matchee, the principal suspect in the case, was removed from the military lock-up in the Airborne Regiment's compound, where he had allegedly tried to hang himself. Jeff had been forbidden to photograph Matchee's removal. I wasn't the only one who suspected a murder attempt to keep Matchee from incriminating senior officers, one of whom apparently offered a case of champagne to the first soldier to kill a "nigger."

"You don't lock a guy up with his bootlaces," I argued, "unless you want him to use them. Besides, the lock-up was too small; he would have had to hold his feet off the ground." It seemed obvious to me that senior officers had egg on their faces, if not blood on their hands.

Jeff was not convinced. "Not bootlaces. Matchee used the cord from sandbags. It was 140 degrees in the shade. He was taking steroids when he killed Arone; they make you aggressive. He was also drinking alcohol and on mefloquine, a malaria medication that is known to cause psychotic reactions in some users. It was common procedure to rough up prisoners. More than one soldier said to me, 'We put the boots to these guys and then throw them out.' Matchee just got carried away."

I had asked Jeff if he thought there was any significance to the fact that Matchee and Private Kyle Brown, another of the accused, were both First Nations soldiers, perhaps acting out the violence and racism that had been inflicted on them back home. Jeff had no theory on that score, but had overheard officers discussing the situation.

"Everybody in Ottawa knows," one of them said, "but it's being hushed up."

"Sure, but part of the problem is we've got Indians involved here."

Then the officers' focus had shifted to the media: "We don't want any of these guys knowing what's going on here." A senior officer threatened Jeff and told him to back off: "If you don't stop asking those kinds of questions, I'm not going to tell you where the landmines are."

"Do you think the disbanding of the regiment was an overreaction, then?"

Jeff looked up from the parcel he was unwrapping. "Yes. Our soldiers in Somalia were distributing food, building a school, a hospital, bridges. They had a strong, effective presence. Here, look at this." He handed me a leather-bound book of photographs.

It was not difficult to see from the album how foreign troops might have been unprepared, out of their element, in Somalia in 1993, and why their presence had provoked fear and suspicion: the sick child being treated by a shaman, his mother pouring cow's blood over his entire body; the soldier in an armoured vehicle wearing a cloth mask against the dust and huge, shaded goggles, resembling a buglike alien;

the boy carving meat from a dead camel's neck, the animal's head still attached, one eye open and unseeing; the lance corporal displaying a scorpion on the end of his knife blade, while a pet monkey inspected his chain and dog-tag; the old nomad with his loose robes, headscarf and thin pole carrying nothing but a metal teakettle on his back; two soldiers showing magazine photos of naked women to young Somali boys; a barefooted Somali fervently praying to Mecca on a piece of cloth beside his truck in the middle of the desert.

As our car approached the coast, I recalled American journalist Robert Paul Jordan's description of the Horn of Africa as harsh and inhospitable. "High arid country mostly—a savannah of acacias, patches of grass, thorny shrubs, tall ant-hills and rocks. When the scanty rains fall, it runs cruel. Then, sheep and goats slowly die. The barrens are strewn with their carcasses." Although my journey overland to Berbera had produced no carcasses, it was never short of interesting geological formations, desolate roadside settlements, baboons and abandoned military equipment from past wars, including the shell of a tank, its barrel pointing a rusty, accusing finger at some unknown target in the sky. The spectacle was intriguing in spite of the distraction of Ahmed Keyse trying to convert me to Islam. He could not comprehend how anyone could doubt the existence of Allah. His tireless evangelism ceased only once, when he paused to point out that the small hills to the east resembled a woman's breasts.

When the Land Rover came to a stop on the white sand, we all piled out, stripped down to our underwear and splashed about in the surf, the driver's and the guards' AK-47s left resting on the heap of clothing a hundred feet from the water, their intersecting barrels forming a perfect X. Not exactly at the ready should there have been an attack. I was grateful that both Allah and security were temporarily forgotten during this ritual baptism in the Gulf of Aden. Our swim was followed by lunch at a sketchy seafood restaurant on the edge of town, where local cats darted in and out between our legs or sat hissing under the table. Offshore, two half-submerged ships served as reminders of past wars. A crowlike bird with a brown body and black wings alighted on the wire fence separating us from the beach, where three young boys were preparing a homemade crab trap. A metal saucepan containing a rock and a bit of bait, with a hole in the plastic lid that made it resemble a bedpan, was submerged and lowered to the sandy bottom.

On the way back to Hargeisa, we passed again through a number of security checkpoints—a simple rope with an unarmed guard asleep nearby—and then made a roadside prayer stop. While my companions did their ablutions, rubbing feet and hands clean with sand, and offered up their prayers on small mats, I studied the harsh, beautiful landscape. What at first I took to be eroded sandstone formations weathered into fantastical, abstract shapes that looked like Henry Moore sculptures—weird family groupings or surrealistic cathedrals—turned out to be termite mounds. These grass-eating termites were amazing architects that built structures resistant to climate change, with controlled ventilation systems and nutrient supplies to last through the winter months. Not the nasty house-devouring pests of my own experience, Somaliland's termites are regarded as conservationists, building up soil in low-lying areas susceptible to flooding so that vegetation can regenerate.

Perspective, I thought. That's what it's all about.

AFTER I BEGAN to read about Somali history and culture, I learned, to my surprise, that Somalia was known as the "nation of poets." The phrase was coined by Richard Burton, the nineteenth-century explorer who wrote in *First Footsteps in East Africa*, "The country teems with poets. Every man has his recognized position in literature as accurately defined as though he had been reviewed in a century of magazines—the fine ear of this people causing them to take the greatest pleasure in harmonious sounds of poetic expressions. Every chief in the country must have a panegyric to be sung by his clan, and the great patronize light literature by keeping a poet." Since Somalia was without a written language until 1972, all communications—from love letters to history, from denunciations and diatribes to political propaganda—were in the form of highly stylized, alliterative oral poetry that was memorized and circulated widely. There was hardly a camel herder, shopkeeper or craftsperson who could not quote proudly the most famous of these poems. According to former Somali prime minister Abdirashid Ali Shermarke, "Our misfortunes do not stem from the unproductiveness of the soil, nor from a lack of mineral wealth. These limitations on our material well-being were accepted and compensated for by our forefathers from whom we inherited, among other things, a spiritual and cultural prosperity of inestimable value. The teaching of Islam on the one hand and lyric poetry on the other."

When I asked Boobe whether Somali poets were still considered important and were still addressing themselves to social and political issues, he was quick to reassure me: "Yes, all topics and issues are tackled in poetry. Poverty, environment, land degradation, AIDS, female genital mutilation, self-reliance, all are tackled. Recent modern songs with music criticize politicians. Ahmed Aw Geeddi, in one of his musical songs, says, 'Garo stop hitting Giiro and Garbo-cas, and you, Giiro, you are older and you should stop the complaint and hitting the others.' Garo, Garbo-cas and Giiro are the names of goats. It is sarcastic that the author used traditional folkloric songs to tackle a modern political issue. The names of the goats stand for the current political leaders." Aw Geeddi, whom I had heard reading his work to an enthusiastic audience at the camel celebration, had paid a price for his satire—boycott by government-funded media.

I pointed out that the government had thrown journalists in jail and wondered if this did not inhibit poets from speaking out. "Poets must have some political poetry; otherwise, they will lose the admiration of the public," Boobe told me. "The government may not like their criticism, but they are seldom jailed." I mentioned the Russian poet Andrei Voznesensky confiding to me during the Cold War, on one of his visits to the "outside," that poets were the unofficial opposition in the old Soviet Union, writing obliquely, drawing huge audiences and in constant danger of having their poems interpreted wrongly—or worse, correctly.

It's impossible not to love a people for whom poetry and camels are top priorities. And as with their camels, Somalis have a number of unique designations for poets. The *crème de la crème* of poets is *af-maal*, the one who has a mouth of wealth. Close behind in prestige is the poet who makes up in facility what he lacks in fame; he is *af-tahan*, blessed with a generous mouth. A lower rank amongst the poets, but still prized for his biting satire, is *af-mishaar*, the poet whose mouth is like a saw and can rip his opponents to shreds. At the bottom of the heap, and the butt of much humour, is *af-garooc*, the poet who is inarticulate or has a deformed mouth.

I had encountered all four categories of poet on the page, in translation; and I hoped to hear readings by poets in the top three categories on my second-last evening in Somaliland. I was invited to the launch at the Hotel Mansour of Boobe's anthology *Deelleey*, a celebration of a famous collaborative chain of sixty-seven poems, initiated by Gaarriye with responses by fifty-one poets, that is regarded as the longest Somali poem-chain and is said to have contributed to the downfall of the Siad Barre regime. When the poets entered the hall, the audience of three hundred stood up, cheering and clapping in unison. This continued for several minutes, the aisles jammed with fans who wanted to shake the poets' hands. Boobe was first to speak: "Tonight, I am extremely happy because the burden of a twenty-seven-year odyssey has finally come to an end. The hardest part was finding the background history of the poets who had composed the series of poems." Ten of the contributors, including Somaliland's two most famous poets, Mohamed Ibrahim Warsame (Hadraawi) and Mohamed Hashi Dhama (Gaarriye), were present to read and comment on both the process and the conditions that prompted them to write *Deelleey*.

Much like the Anglo-Saxons, Somali poets write in a highly alliterative style, delighting in the rapid fire of recurring vowels or consonants. On my way back from Berbera, to avoid the subject of religion, I had tried to illustrate the technique to Ahmed Keyse by constructing a mundane six-line alliterative parody:

> I would like to write a deelleey
> daily, duly, dearly, but never dully.

There has been no dearth of death,
doubt not, O dutiful dromedary,
descending the Dervish dunes,
the drafts, the dictatorial divides.

Ahmed Keyse was neither impressed nor amused, but I had proven my rank as *af-garooc*, he of the deformed mouth. Although we were hurtling southwest across the desert with our two armed guards, he continued to urge my soul northeast in the direction of Mecca.

Deelleey had helped to launch the armed struggle against the brutalities of the Siad Barre regime, but Hadraawi reminded his listeners that not all contributors were of one accord about the alternatives. Siad Barre had not followed Plato's example and banned poets from the ideal republic. Instead, he had courted them, tried to enlist their help; only when they refused to comply did he persecute them. Hadraawi's five years in jail had been prompted by a poem called "The She Camel," which depicted Somalia as a slaughtered female camel fought over by numerous contenders, each of whom carves, cooks and consumes a chunk, so that even bones and skin disappear in the hot, dry climate. In lines from another Hadraawi poem, "Has Love Been Blood-Written," translated by Martin Orwin, the speaker links the private and the public selves:

if self sacrifice is not made
the breath of life not exchanged
if one does not wait
for an enduring legacy
the building of a house upright
children and earthly sustenance
then the kisses and intentions
are nothing but superficial
a poison sipped to satisfaction
in that one same moment
like hyenas snatching

a girl of good repute
as they hide themselves
in the higlo tree
to pounce out quickly
each man is expectant
for what will fall to him
a hyena and his grave hole
the honour he has trampled
the modesty he has snatched
the lying illusion
this does society harm

After he was released from prison, Hadraawi spent the early 1990s in exile in the United Kingdom, but could not escape the sense of being an outsider. As he explained in an interview, "I felt I had lost my way and I decided to return back home where I am one of the most respected people." Home again, he not only worked as a teacher of literature and mentor for a new generation of writers but also led a peace march from one end of Somalia to the other. Although he resides in the new Republic of Somaliland, Hadraawi's fame and peace-building efforts at home and in the diaspora had made him much beloved by all Somalis.

I sat close to the front of the hall with Martin Orwin and British poet and editor W.N. (Bill) Herbert, who teaches at Newcastle University and had come to Somaliland to work with Martin on translations of Gaarriye's work. Martin would do the rough translations; Bill, who did not speak Somali, would work on polishing the English versions. Martin, already known to the audience, and Bill and I had been introduced and welcomed earlier. As one of the younger poets belted out his short lyrics, I glanced at Bill. He looked as bemused as I felt. The music of foreign words swirled around us, nesting in our ears. Then Gaarriye took his turn at the podium.

"Tonight it's as if I'm born again," Gaarriye announced, immediately galvanizing the audience with his wit, charisma and colourful costume, the highlight of which was a blue, orange and white *macawis* with narrow triangular designs that looked like flames. He was constantly in motion; half the photos I took would turn out to be blurred. For the second time in a week, I was struck by the patience and goodwill of the audience, attentive and enthralled, sitting through four hours of poetry. Gaarriye kept them glued to their seats. I could not understand his *deel-leey* poems, as they were read in Somali, but the cadence and recurring alliterative elements could not be mistaken. However, I knew several of Gaarriye's poems that Martin had translated literally and David Harsent had polished, including this piece about the life and struggles of Nelson Mandela:

> This poem is a gun.
> This poem's an assassin.
> Images mob my mind . . .
> This pen's a spear, a knife,
> A branding-iron, an arrow
> Tipped with righteous anger.
> It writes with blood and bile.
>
> I take this bitter ink,
> Blood-red, to make my mark;
> Corruption from the wound,
> Sap from the poison-tree,
> Aloe and gall and myrrh.
>
> This poem's a loaded gun,
> This verse a Kalashnikov.
> I aim it at the snake
> That slithers to our children
> And strikes! See where the tell-tale
> Blood-beads pearl on the skin.
> The snake, the Prosecutor,
> The Oppressor, the Judge, the Jury —
> You must always aim for the head.

From the few snippets Martin translated for me, it seemed as if my own head was the target of much of what was said: the eternally interfering foreigners, the blight of Africa.

One of my favourite Gaarriye pieces was "A to Z," a celebration of the creation of a written language, in one of Martin's finest translations:

> Caalin, listen, I'm going to travel
> From A to Z carried by language—
> The alphabet, alive on the page.
>
> I write the words and send them to you;
> You sing to the wind, and the crows as they fly
> Carry my lines through the noonday sky
> Chanting each to each. The ants
> Become orators. The gossiping camels
> Crowd the waterhole, eager for rumours.

The poem covers a lot of ground, linguistic and historical, lamenting what has been lost in the culture, but ending on this triumphal note:

> Enough! I've written all that I need
> To write, except to praise the men
> Who talked the language into being—
> Statesmen, thinkers, poets, who gave
> Somali poets a new way with words.
> We could raise a statue to them and set it
> Above the image of Jupiter . . .
> Or perhaps we should honour them in poems
> That use all the letters from A to Z.

I'd met both Haadrawi and Gaarriye, but there would be no opportunity for extended conversations or interviews. Over coffee earlier in the day, Gaarriye had lamented that I was leaving so soon, as he'd planned to take me with Martin and Bill on a brief tour of Somaliland, visiting communities where he was giving readings and talks. I was disappointed not to have been warned in time to make other arrangements. We promised to meet again, either in Hargeisa or in Canada.

IN MY TRAVELS across sub-Saharan Africa, I'd come to understand there were no easy solutions for protecting human rights. Notions of justice covered the spectrum from revenge and punishment to forgiveness and reconciliation. But there were two constants—the need for peace and the need for healing, however slow the process. I'd encountered various types of healers and seen the healing effects of good work. I'd seen, too, how the arts provided comfort and inspiration to downtrodden spirits, to people who had lost everything except the capacity to breathe.

Now, it seemed, I'd found a society, its roots deep in the past, where poetry was both a healing potion and a political force. And yet questions about the use and misuse of art and rhetoric, old concerns of mine that had surfaced again in the question period after my lecture at the University of Addis Ababa, still hung in the air. Journalist Lark Ellen Gould recounts an instance of the negative use of poetry: "Somalis tell the story of an Ogaden chief who went to the tree of an enemy tribe and offered 114 points of introduction and then another 114 points of argument, in honor of the number of chapters in the Qur'an. According to the legend, the chief spent days convincing the tribe that a negotiated peace was at hand, while actually keeping the enemy spellbound, buying time for his approaching warriors. By the hundredth point of argument, two days later—a performance played entirely from memory—he had lulled the opposing tribe to sleep; his army arrived and slew them."

When I asked Martin Orwin about the quality of Somali poetry, his response was measured. "The last two decades in the Horn of Africa have been times of great upheaval, culminating in the early '90s in horrific violence in some parts of Somalia and the consequent displacement of a great number of people throughout the world. Much of the poetry that has been written over these years has been concerned with this and some imaginative and powerful poems have been composed. As has been the case throughout Somali history, some of this poetry is partisan, supporting or denigrating according to the allegiance of the poet. Poetry which becomes most widely known, however, tends to be that which deals with the situation as a whole and speaks to a wider section of society."

In the epics of Homer and Virgil, the gods are afflicted with problems similar to ours—ego, lust, greed, ambition—all of which lead to trouble, even to war. For the Greeks, as for the Somalis, poetry was one of the principal means of communication, capable of accommodating politics, history and psychology. Towards the end of an unsuccessful ten-year siege to recapture the incomparable Helen of Troy, whose beauty launched a thousand ships, the Greeks build an enormous wooden horse and leave it outside the gates of Troy. The Trojans, believing the war over and the horse a departing gift from the enemy, bring the huge wheeled toy into the city and drunkenly celebrate the end of the siege. During the night, a trap door in the belly of the horse opens and an elite guard led by Odysseus drops to the ground, opens the gates to the army and proceeds to sack the city.

Homer wrote about these matters in his poetry for the same reason that Hadraawi and Gaarriye do: because his people had lost sight of civic virtues, because violence and war threatened to engulf the nation and destroy the fragile social fabric. He wrote about them because he believed the lessons contained therein would help Greeks (and still help the rest of us) to understand destructive impulses and cultivate more sensitive forms of government and a more enduring social contract. Poetry is a primary process language with the power to touch us at the deepest levels, a kind of Trojan Horse, a subversive force that gets under the skin, into the bones; it moves at great speed below the radar, outside the usual paths of rational thought. This network of sounds we learn as infants, which serves as a substitute umbilical cord, persists as a sort of organ based deep in the psyche, a rich, primal language of the heart.

As Boobe's book launch concluded, the audience swarmed the poets to shake hands and congratulate them. Gaarriye, his arm around the shoulders of an animated fan, laughed uproariously as cameras flashed. Poetry and peace—I wondered if there was a connection to be made. Poetry is one of the healing arts; it speaks to the wound in each of us, to the part of us that is damaged, incomplete, so often in ruins. For Dylan Thomas, who struggled with alcoholism and other demons, poetry was therapeutic. "Out of the inevitable conflict of images," he wrote, "—inevitable because of the creative, destructive and contradictory nature of the motivating centre, the womb of war—I try to make that momentary peace which is the poem." The peace that poetry affords is not permanent; it is no more enduring than the peace afforded to Tutsis and Hutus, to Ethiopians and Eritreans or to Israelis and Palestinians. It needs to be renegotiated daily. Like peace itself, poetry is a process, a way of life, a daily discipline of self-assessment, commitment and renewal, driven by good faith. Language, like politics, is an unstable medium; it's dirty with sound, with secondary meaning and with unexpected connotations that come zinging in from the side to torpedo the best of intentions. Gaarriye and Hadraawi have learned to work with this unstable medium, bring it under control, make it sing and generate hope. In the hands of such talented writers, poetry has the power to touch what is broken in us, to mend what is damaged. As the poets, feminists and psychologists tell us, the place of damage is also the place of power.

Back at the Oriental Hotel, I was still pondering these matters. One of the things I had been striving towards on my sub-Saharan journey, in addition to some understanding of Third World realities, was a revolutionary "poetics" that moved beyond the impersonal voice of the all-knowing observer to the restive, deeply troubled voice of the learner, whose ignorance places him or her continually at risk. As Irish poet Eavan Boland so succinctly explains, "I do not believe the political poem can be written with truth and effect unless the self who writes that poem—a self in which sexuality must be a factor—is seen to be in radical relation to the ratio of power to powerlessness with which the political poem is concerned." As Boland—a woman in a university and a male-dominated world—knows all too well, every genuine exercise and utterance must be performed in the full knowledge of the agent's position in a hierarchy of privilege and power. For me to talk about human rights and injustice or poetry as a healing art in Africa is to be constantly reminded that I live in a society where poverty and violence, though serious and increasing, are still the exception rather than the rule, and where the making of a work of art is only political insofar as it acknowledges and struggles against its own radical uselessness.

Somali culture has also produced distinct voices in the diaspora, including those of novelist Nuruddin Farah and rapper K'naan. Farah's first novel, *From a Broken Rib*, explores the tumultuous life of an intelligent country girl named Elba who has an independent mind and must learn to navigate the labyrinths and double standards of a male-dominated society. Without malice, she elects to play by the same rules in this unfair world, where it's obvious, as her friend Asha concludes, "men really ought to be beaten." In his subsequent novels, Farah does not hesitate to address the issues Boobe mentioned: violence, poverty, female circumcision and political corruption. K'naan, who grew up in Mogadishu during the civil war and now lives in Canada, has been no less controversial, gaining an international reputation for his music and lyrics and for speaking out against the greed, violence and bloodshed that forced his family from Somalia and the failure of the international community in its response. On his first CD, *The Dusty Foot Philosopher*, which combines African rhythms with rap and hip-hop verbal strategies, he ranges over everything from the problems of African slums to the struggles of immigrants in First World ghettos. It did not surprise me to learn that K'naan's strength lies in an early love of words: "Seven, eight—you're a pretty established human being when you're that age in Somalia," he said in an interview. "Everyone is expected to be a linguist, even little children. For us, language is what informs the universe. Your articulation is your manhood." In addition to a love of language, Farah and K'naan share the vitality, irony and independence of spirit that I found so refreshingly evident in the poets of Somaliland.

While Human Rights Watch has declared Somalia a "shell-shocked" disaster case, where thousands of innocent civilians have been killed and close to a million displaced or driven into exile, the Republic of Somaliland, this vibrant and relatively successful young state, dreamt into existence by freedom fighters and poets who have turned in their weapons and taken up the healing arts, remains a safe haven, for which international recognition and heightened support cannot come soon enough. Whereas termites are considered pests in North America and the writing of poetry a barely tolerable affectation, here both have an important function. As I prepared to leave Hargeisa and make my way home via Djibouti, Addis Ababa and London, I could not help wondering what might have happened in the Horn of Africa had Canada and the United States sent a contingent of poets to Somalia instead of armed forces.

Boobe, whose passion for Somaliland is rivalled only by his love of poetry, was at the airport to see me off, bearing gifts: a wooden camel bell and two beautifully carved wooden spoons, as if to say, once again, like the tapping of knuckles on prison walls, I am with you.

eight *Walking Wounded*

AFTER ALL this, what do I know? The taste of shame, a sour residue not easily washed away. I have seen children foraging in open sewers; a church where bodies had been piled four deep, a place no self-respecting communion wafer would be found dead in, but where the Holy Ghost is still said to reside. I have seen camels, a mother and foal, so awkwardly elegant, about to step onto the desert road until my motorized approach whispered caution; a red-eyed baboon casting aside a banana peel, whose three strips of skin resembled an under-endowed starfish in the dust. I have seen white clouds laid out like stepping stones above a smoking volcano, signalling escape, while below, on a jungle road, a family of four fling themselves into the bed of a fleeing pickup, the small brown goat they were carrying left bewildered in the dissipating diesel fumes. I have seen beauty and suffering, victims enough to swamp the shallow vessel of my concern, lined up outside my interview room, speaking twenty languages and waiting for a word, not just any, but the one word that will give them hope.

I've seen signs of agency, too. A small man badgering me in French, arguing a future for his eldest son; a woman raped nine times, her ruptured fistula scarcely mended, talking of home—not my word for that place of ignominy, of pain—where she'll care for seven orphaned nieces and nephews, casualties of gold, of coltan in my cellphone. I've seen the pink scars where facial features used to be, like smears of cheap lipstick waiting to be kissed, this time by the plastic surgeon's scalpel. I've seen images of journalists, teachers, lawyers, fixers, teenage girls, members of civil society, facing down rapists, murderers and militiamen, whose dirty fingers caress the safety catches of their AK-47s. Yes, and the escaping sigh that ought to have ended in despair or death, but was drawn back into the lungs as fuel for the struggle—Lumumba's dream rising from the acid bath his bones dissolved in—a man and a woman, their lives on the line, smooth the creases of an unfolded map spread out on the weathered, wooden desk to show me the areas of conflict and who is to blame. Heroic struggles, large and small, against microbes, viruses, kleptocrats, CEOs, the World Bank, the IMF, people like me, well-meaning, dangerous, armed with the wrong ideas.

The Polish writer Ryszard Kapuściński, who came to know Africa as well as any Western journalist, once observed that a journey "neither begins in the instant we set out, nor ends when we have reached our doorstep again. It starts much earlier and is really never over, because the film of memory continues running on inside of us long after we have come to a physical standstill." In the months following my return, there was no question of forgetting sub-Saharan Africa. I deliberated over my portrayal of events and people with uncharacteristic urgency, even fury. And while I wrote, other levels of activity unfolded. I fired off letters to various authorities, offered financial help when I could, kept in touch.

As I walked along the shoreline in a light drizzle, a cool west wind from the Strait of Juan de Fuca tugging at my raingear, I could visualize my Rwandan friends, inching slowly uphill by mini-bus or on the back of a motorcycle through contending traffic and the cacophony of horns, circumnavigating the roundabout formed by the Place de l'Unité National, stepping off a bus into heat, diesel fumes and congestion. Not for them casual shopping or a latte at the Bourbon Street Café and Internet Bar.

Valentin's son Luka, I learned, had graduated from high school and needed help to continue his studies. Sam Nkurunziza was enjoying his work so much he had abandoned both Uganda and medicine for the quick cure of Rwandan journalism, though he continued to take courses towards a degree in arts and communications. His article about Marcel Gatsinzi, the Rwandan minister of defence accused of complicity in the genocide, was never printed. The Gacaca court declared itself incompetent to judge and referred the case to the criminal courts. The International Criminal Tribunal for Rwanda remains ponderously slow in its deliberations, but other jurisdictions have begun to respond to the call for justice in relation to the genocide. The Quebec Superior Court in Montreal found Rwandan Désiré Munyaneza, the son of a wealthy Butare businessman, guilty of seven counts of genocide, war crimes and crimes against humanity and sentenced him to twenty-five years with no possibility of parole. Germany arrested accused *génocidaires* Ignace Murwanashyaka and Straton Musoni, president and vice-president respectively of the FDLR, even as those forces continue their predations and retake positions deserted earlier in DR Congo.

Since my time there, Rwanda has been admitted to the Commonwealth for its achievements and reviled by human rights groups and the legal profession for its draconian "genocide ideology" bill, which seems to entitle the government to arrest anyone who even thinks about the matter of genocide, never mind questioning the official narrative. The country is also under fire for increasing nepotism. Following grenade attacks in downtown Kigali and the arrest of opposition politicians in 2010, Paul Rusesabagina, owner of the Hôtel des Mille Collines, wrote two open letters. In the first, addressed to U.S. president Barack Obama, he asks for a truth and reconciliation process in the Great Lakes region; he is highly critical of Kagame for initiating the Rwanda conflict with the first rebel attack and claims that children with two Hutu parents cannot get free education in Rwanda. His second letter was to French president Nicolas Sarkozy, advising him that Madame Habyarimana, wife of Rwanda's assassinated president, could not possibly get a fair trial in Rwanda; so, if there are sufficient grounds to try her, he advised that she be sent to the Rwanda tribunal in Arusha. While the notion that Kagame's rebel attacks in 1990 initiated Tutsi-Hutu struggles is grossly inaccurate, these two letters from a hero of the genocide could prove explosive and a PR stumbling block for the Rwandan government.

In the fall of 2009, Gilbert, the artist I so enjoyed meeting in Gisenye, suddenly found himself in prison. At age fifty-six, he had been called as a witness in the trial of several men accused of involvement in the genocide, but his testimony somehow angered the judges, or caught them off guard, and he was thrown in jail. Although he was released after several months, the government stopped his children's educational assistance during his detention. The details of Gilbert's story remain somewhat vague, but the fiasco is another indication of the tenuousness of Rwandan justice. In a much-publicized news report, one of Kagame's aides declared that multi-party politics and an unfettered media are enemies of the new Rwanda, a statement certain to trouble investors and supporters who like to tout the fledgling regime as a democratic model for Africa. While healing continues in many sectors of Rwandan society, including the area of women's rights, thousands of people still languish in prison, with justice a long way off.

Media and Internet reports from Uganda, the Great Lakes and the sub-Saharan region arrived daily in my electronic mailbox, reminders that the region was constantly in flux. Although less violent than DR Congo, no place seemed more volatile and mercurial than Uganda, as if its turbulent history was a legacy that lurked always just under the surface. I was greatly concerned about how this might be affecting the activists, health care workers and other angels of mercy I'd met. Headlines announced the discovery of multi-billion-barrel oil reserves in the Lake Albert region on the Uganda/DRC border and tracked a bill being debated in the Ugandan Parliament that would criminalize homosexuality. The American evangelicals behind the homophobia and hate-mongering have strong local support, which could nudge Uganda in the direction of a police state. Some claim the bill, widely condemned by other nations, human rights organizations and even the Vatican, was intended to draw attention away from corruption and the government's efforts to garner support through tribal alliances in the buildup to multi-party elections. Whatever the motive, that such a regressive bill was introduced in Parliament suggests a further hardening of the political arteries in Museveni's clique, not to mention the growing influence of hardline Christian fundamentalism.

To my relief, Victor Ochen and his friends at African Youth Initiative Network were able to take little Michael to Mulago Hospital in Kampala to have his recurring leg infection examined by experts. Doctors there diagnosed cancer and recommended amputation, but Victor was not convinced. He redoubled his efforts, locating a surgeon who dismissed the cancer diagnosis, opened the wound and cleaned the thigh bone of all infection. In May 2010, a high-profile football game was organized by AYINET and an NGO called No Peace without Justice to bring public awareness to the plight of war crimes victims. The game, held in Kampala, was planned to coincide with meetings of the International Criminal Court with its States Parties, the first major review of the organization since the Rome documents were signed. Through Victor's efforts and diplomacy, Michael was selected to receive the dignitaries and to take the ceremonial walk hand in hand with Ugandan president Yoweri Museveni and the UN's secretary general, Ban Ki-moon, both of whom had donned uniforms and played football alongside war crimes victims, members of Uganda's national team and other States Parties.

Chief Prosecutor Luis Moreno-Ocampo addressed the ICC review conference in Kampala with these words: "There is a need for consistency. Massive crimes require a careful plan. Certainty that these crimes will be investigated and prosecuted will modify the calculus of the criminals, will deter the crimes, will protect the victims. It is your time for action." Although Moreno-Ocampo was clearly addressing his comments to Uganda, in the hope that the government would intensify its efforts to capture Joseph Kony, fear of prosecution had obviously not proven a deterrent to the LRA. And according to a report published by the International Crisis Group on April 28, 2010, "Not even a complete military victory over the LRA would guarantee an end to insecurity in northern Uganda. To do that, the Kampala government must treat the root causes of trouble in the area from which the LRA sprang more than twenty years ago, namely northern perceptions of economic and political marginalisation, and ensure the social rehabilitation of the north."

Nancy's story, with its emphasis on forgiveness and healing, suggested a quicker and more productive path to the kind of results Moreno-Ocampo envisioned. While she was recovering from the first bout of reconstructive surgery, made possible by the ICC's Trust Fund for Victims, Victor sent an e-mail with photos and a long written testimony by Nancy attached. She talked about how she had resisted offers of help, not believing anything could be done and wanting only to end her pain. "But they insisted and brought me to the hospital where I received lips reconstructions. That was first time I saw something good happen to me, and I eventually got the idea of going back to school. My life changed and I couldn't let 20 minutes go by without looking myself in the mirror. Three months after, Watoto [holistic care program] team approached me that AYINET had given them my profile and I could still have more operations. They took me again to hospital and I receive surgery of the nose and right now I feel I am back, no more stigma and I really feel so good and grateful to all the teams behind my surgeries."

The amazing repair work done by Dutch plastic surgeons prompted a second outburst of joy from Nancy. "I want to continue with my education and be able to support my daughter and my mother who has been so much in pains over me. I have been through very hurting moment in life but I now feel the kind of love from people who are even not related to me that I never felt before. My daughter keep asking me not again remove my lips and nose, and that I now look like other people's mothers and she is very happy right now."

I read reports on the conflict in DR Congo, including the murder of six mountain gorillas in the Virunga National Park, thanks to a deadly mix of displaced persons, contending militias and Congolese government troops that had converged on the area, threatening the habitat and wildlife. Huge tracts of timber were being cut by charcoal traffickers; desperate or indifferent poachers in search of "bush meat" were driving the gorillas towards extinction. U.S. secretary of state Hillary Clinton paid a surprise visit to the HEAL Africa Hospital in Goma shortly after I was there, promising increased pressure to end the use of rape as a weapon of war. Apparently unaware of the heroic work being done by HEAL Africa, she announced a $17-million USAID package to construct another hospital to offer identical services. Nonetheless, her offers of financial aid, police training and continued monitoring boosted the morale of the overwhelmed staff and her words were a temporary clarion call:

> Our commitment to survivors of sexual and gender-based violence did not begin with my visit to Goma, and it will not end with my departure. We are redoubling our efforts to address the fundamental cause of this violence: the fighting that goes on and on in the eastern Congo. We will be taking additional steps at the United Nations and in concert with other nations to bring an end to this conflict. There is an old Congolese proverb that says, "No matter how long the night, the day is sure to come." The day must come when the women of the eastern Congo can walk freely again, to tend their fields, play with their children and collect firewood and water without fear. They live in a region of unrivaled natural beauty and rich resources. They are strong and resilient. They could, if given the opportunity, drive economic and social progress that would make their country both peaceful and prosperous. Working together, we will banish sexual violence into the dark past, where it belongs, and help the Congolese people seize the opportunities of a new day.

Rousing words, noble sentiments. However, recent reports from Human Rights Watch, Oxfam and other NGOs indicate that the LRA's rapes and murders have increased in the north, and the situation has deteriorated again in Rutshuru as FDLR *génocidaires* return to the territories from which they were driven by the joint Congo-Rwanda military operation and take their revenge on the civilian population. The militias and proxy armies continue to pillage and plunder. A brief dip in hostilities farther south, and fickle media attention shifting to other fronts, meant that Célestin was temporarily out of work as a fixer. Amidst a severe economic downturn and fears of a resurgence of armed conflict, he made the disturbing observation in an e-mail that "a hidden hatred is living in people's hearts about all humanitarian organizations based in Goma pretending to assist vulnerable people."

While UN officials pump out reports on the positive effects of the two joint military offensives in the Kivus, it's clear that efforts to integrate rebel militias into the national army have merely increased the number of rogue elements wreaking havoc on the civilian population. In November 2009, Human Rights Watch reported on government soldiers who attacked five hamlets, killing eighty-one civilians. In one of these hamlets, soldiers decapitated four youths, chopped off their arms, then raped sixteen women, killing four of them. Widespread criticism of government forces for human rights abuses prompted a decision to expel UN troops, currently the world's largest deployment of peacekeepers, a development that will not improve conditions on the ground or reduce impunity.

Hillary Clinton made a significant error in her assessment: fighting is not the fundamental cause of violence in the DRC. Greed, homegrown militias and foreign interests operating in an unregulated mining sector are the root causes. In reports from Mining Watch, Amnesty International, the International Peace Information Service and other observers, extractive industries in DRC and elsewhere in Africa face serious criticism, but they have done little to mend their ways. Despite company attempts to downplay the long-term effects of chemical pollutants, for example, cyanide spillages in Ghana are decimating communities by polluting local streams and water tables. But even as companies under scrutiny ignore safety concerns and deny human rights abuses and environmental degradation, grassroots organizations have begun to have an impact. In 2005, after powerful lobbying resulted in bad publicity and a massive divestment of pension funds, Talisman Energy was forced to sell its oil assets in Sudan, which were being used by the government to finance its assault on Darfur. In response to reports of unacceptable ethical and environmental practices at the Porgera Mine in Papua New Guinea, Norway's government pension fund withdrew its investments in Barrick Gold's operations, selling shares worth $229 million. Forcing companies to withdraw is not always the best solution, however, especially as Chinese and Indian interests increasingly step into the breach with even worse human rights and environmental practices.

Today, Barrick's website creates the impression that the company is no longer in business to make money, but to do social work, improve health and education and take care of the environment. King Leopold used a similar strategy. However, in 2010, Barrick Gold threatened legal action against yet another small publisher, this time in British Columbia, over an as yet unpublished manuscript it suspected would be damaging to its reputation. The book, *Imperial Canada Inc.*, was withdrawn from the Talonbooks website, one more nail in the coffin of free speech and opinion. Banro Corporation, another player in both the scramble for gold and the silencing of dissent, has begun construction on its site in Twangiza, not far from Bukavu, where it believes there are reserves of at least 11 million ounces of gold. There is much concern about the social, political and environmental impact of this open-pit project, as well as the deal struck with DRC's government, which, in exchange for a few concessions, including the release of an area of land, gives Banro 100 percent ownership of the gold. While complaining about the craziness of dealing with Congolese bureaucracy, Banro's CEO, Mike Prinsloo, as reported by Geoffrey York in the *Globe and Mail*, is clearly pleased with himself: "Our projects make money at $650 an ounce and they are lucrative at $850. And at $1,050, they are nice projects." Mining companies will clearly not police themselves. Tough regulations and an ombudsman are needed by the industry's home countries, with the power to impose hefty fines and criminal prosecution for polluting, social disruption and human rights abuses.

In Goma and Bukavu, civil society struggles to foster enlightenment and change. In the fall of 2010, 1,700 women marched to protest escalating sexual violence in the DRC, with Olive Lembe Kabila, the smiling wife of President Joseph Kabila, prominent in the photographs. Every woman in the march represented at least ten rapes in the previous year: the UN estimated 17,507 sexual attacks during that period, but since countless rapes go unreported, the real figures would be much higher. Regaining dignity would be easier for women if Ms. Kabila's smile were transformed into positive action at the government level.

Ethiopia was very much with me as I continued to monitor the situation of those I'd met. There is no shortage of money in the country for those in positions of power, most of it coming from foreign aid, but little filters down to the masses. Ryszard Kapuściński lived for a time in Addis Ababa and reported on events for Polish media. His observations in *The Emperor* on the pomp, splendour and sheer indifference to poverty on the part of Haile Selassie and his cronies are startling to read, especially in contrast to the portrait of starving people foraging for scraps outside the palace during a banquet for visiting presidents:

> I noticed that something on the other side was moving, shifting, murmuring, squishing, and smacking its lips. I turned the corner to have a closer look.
>
> In the thick night, a crowd of barefoot beggars stood huddled together. The dishwashers working in the building threw leftovers to them. I watched the crowd devour the scraps, bones, and fish heads with laborious concentration. In the meticulous absorption of this eating there was an almost violent biological abandon—the satisfaction of hunger in anxiety and ecstasy.
>
> From time to time the waiters would get held up, and the flow of dishes would stop. Then the crowd of beggars would relax as though someone had given them the order to stand at ease. People wiped their lips and straightened their muddy and food-stained rags. But soon the stream of dishes would start flowing again—because up there the great hogging, with smacking of lips and slurping, was going on, too—and the crowd would fall again to its blessed and eager labour of feeding.

Resignation and despair are the legacy of this kind of long-term disparity between rich and poor, as Kapuściński illustrates so eloquently. While Ethiopia has endured several regime changes in the last hundred years, at least one purporting to be driven by "the people," nothing resembling a genuine proletarian uprising has taken place to bring about serious change. A hierarchical system has deep roots, not easily removed. Those in power change, but the system and the attitudes that sustain it do not.

Ephrem, the Eritrean refugee who suffered torture and imprisonment in both his own country and Ethiopia, was accepted for resettlement by Australia and sends me regular reports about his new life there, occasionally addressing me with the phrase "G'day, mate." Although trained as a professional psychologist, he is volunteering as a translator, is taking driving lessons and has registered for a two-year diploma in nursing, in the event that he is not accredited by the Australian Psychological Society. Yusuf Warsame, my guide in the community of Somali refugees in Addis Ababa, whose file had dropped off the Department of Citizenship and Immigration radar, has finally, after various setbacks and formalities, been relocated with his family to Canada. As I write, they are in North Battleford, Saskatchewan, suiting up for winter in one of our coldest cities. His children are in school and he is a happy man, working—I'm not surprised—as a consultant for new immigrants. In the lives of many refugees I met through Yusuf's work in Addis, small shifts seem to have happened. Halima found a safe place to live and have her baby; Faduma and her son were re-registered as refugees after a decade in limbo; Hossein, the diminutive refugee from Mogadishu who dreamt of becoming king of Sweden, has been given a new wardrobe and reunited with his aunt in Nairobi. I don't know if the theatre she manages is a cinema or a playhouse, but either should suffice to stoke the imagination of this courageous and resilient child. Chang Ter Pusch, the Sudanese refugee I met at the Jesuit Refugee Centre, has been to see an eye specialist. And the two brothers who fled Congo are slated to take courses in electricity, refrigeration and music if they can be rallied from lethargy and deep depression. Drops in a vast bucket, of course, but significant drops.

Kemer Yousef, the Oromo singer who fled Ethiopia for his life, released his latest CD, *Nabek* (2008), to international recognition. Forgiving his old enemies, he returned triumphant to Addis Ababa to do a six-concert tour, sponsored in part by the unpopular government, which seems keen to make political hay out of its popular son. Oromo legend Ali Birra, for his part, was offered an honorary degree by Jimma University on June 9, 2010. Whether this marks a shift in attitude towards greater recognition of the Oromo and other persecuted minorities in Ethiopia remains to be seen. The news has been encouraging, but the streets of Addis Ababa still swarm with refugees, trauma victims who cannot find employment and who go to bed, if they have one, desperate and hungry. Although recalcitrant and determined to hold onto power, the government cares about foreign aid and, in response to international criticism and pressure, released pop star Teddy Afro from prison in August 2009 and opposition leader Birtukan Midekssa in October 2010.

The small island I live on is blessed with an unusually high rainfall, something desperately needed by those trying to survive in the Horn of Africa barrens. Oceans here are being polluted by toxic waste and industrial fertilizers. The fish, so plentiful when I was a boy pulling in a salmon-laden gillnet with my father in the deep fjords of the north coast, are under severe threat of extinction by overfishing, climate change and invasive farmed species. Waters in the Gulf of Aden, as they prepare to meet the larger body of the Indian Ocean, are no different, carrying chemicals, pollution from shipping in the Suez Canal and the Red Sea, oil spills and sewage from Djibouti and Yemen. Local history has operated in the same overwhelming way, with wave upon wave of invaders, including Phoenicians, Arabs, Chinese, Europeans and North Americans. Although they carry mental and genetic baggage from the past, Somalis have managed to forge a unique culture. You can see the racial mixture in their tall, slender bodies and the bone structure of their dark, almost Caucasian faces; you can hear the music and love of language in their speech and poetry. I marvel at the resilience with which they have resisted some foreign influences and adapted others to their independent natures and nomadic lifestyle, developing a more tolerant and less doctrinaire brand of Islam.

Other than the freeing of Canadian journalist Amanda Lindhout for a hefty ransom, news from the Horn of Africa has not been promising. The World Food Programme has pulled out of areas in southern Somalia because of increased fighting and security concerns, leaving more than a quarter of a million displaced persons without food, medicine or refuge. As with the Taliban in Afghanistan, the deeper the armed conflict in Somalia the greater the likelihood that Islamic fundamentalists will win the day, working to spread their radical theology north and west. In the tiny, precarious Republic of Somaliland, all seems quiet except for the camels, the poets and the electoral commission. Delayed almost two years by technical challenges and foreign interference, the presidential election was held on June 26, 2010, a historic day exactly fifty years after Somaliland first gained independence from Britain. On July 1, opposition leader Ahmed M. Mahamud Silanyo was sworn in with a majority and a peaceful transition of power took place. Six months after my departure from Hargeisa, I received the first draft of Hassan Madar's anthology of Somali poems about human rights, a rich bouquet of this unique people's concerns and aspirations.

In his book *Becoming Somaliland*, Mark Bradbury examines the history of this new and relatively stable political entity and discusses its long-term viability as a state. He makes the important but too often ignored point that Somaliland is more embracing of fairness and liberty than are many of its African and Gulf neighbours. Bradbury's notion of state building as a slow, methodical and ongoing process, a marriage of consensus and trust, strikes me as crucial. Poetry alone will not guarantee the success and survival of Somaliland, but it has in common with that other work-in-progress, the state, a commitment to justice and healing that measures success step by step, word by word, and is always subject to revision.

At the International Criminal Court, the wheels of justice turn slowly. Thomas Lubanga Dyilo's trial, which had been in danger of being thrown out because the prosecution withheld evidence considered exculpatory, nears completion and the participants are preparing their closing briefs. The results, whatever the verdict, will be both telling and contested. So far, Omar al-Bashir has resisted arrest, and Joseph Kony and other high-profile accused are still at large. Whether their sleep has been disturbed by the court's charges, as the UN's Béatrice Le Fraper du Hellen predicted, I can't confirm. Chief Prosecutor Luis Moreno-Ocampo continues to give more speeches than a pre-election politician, as if his fire and momentum might make up for the astronomical costs and interminable delays in the system. It's still the case that selective tribunals too often serve only the interests of the great powers, but I am pleased to learn that the ICC's victims' reparations program, which funded the African Youth Initiative Network and other important programs, has been expanded and that justice and healing are increasingly seen as intertwined.

Foreigners in Africa are likely to misinterpret most of what they see. Barbara Kingsolver, in *The Poisonwood Bible*, puts the matter cleverly: "Poor Africa. No other continent has endured such an unspeakably bizarre combination of foreign thievery and foreign goodwill." Sadly, the goodwill may have damaged Africa even more than the thievery. I can identify with her character Adah, the missionary's daughter whose physical disability allowed her, like Emily Dickinson, to see the world slant and to describe herself as "a crooked little person trying to tell the truth. The power is in the balance: we are our injuries, as much as we are our successes."

Although justice takes many forms, the scales in sub-Saharan Africa tip towards restoration, not retribution. Simon Ntare may be right that shame is more difficult to overcome than grief. The dust of all countries is tainted, not least my own, and the scramble for loot in Africa is ongoing. Writing for the *Observer*, John Vidal reports that foreign interests are also buying or acquiring ninety-nine-year leases on huge tracts of farmland in Sudan, Ethiopia, Congo and a host of other

African countries—land grabs already surpassing in volume the entire land mass of the United Kingdom. This new agribusiness is under way not to benefit hungry Africans, but to grow biofuels, test genetically modified crops and ensure into the future high profits for corporations and cheap food supplies for Saudi Arabia, Israel, China and other wealthy nations. The displacement of subsistence farmers and competition for scarce water resources—one foreign-owned agricultural project, Vidal suggests, uses as much water per day as 100,000 Ethiopians—are certain to create new hardships and ignite conflicts in Africa.

The forces of greed and bad governance, and the speed with which news of the latest travesties reaches us through the media, continue to haunt me, as does the endless procession of displaced persons staggering along muddy roads, wild-eyed with fear or desperate with hunger, clutching a kettle, a few twigs of firewood, a listless child. But I take comfort in remembering the work being done by amazing African women and men—Peninah, Virginie, Victor, Kubuya, Yusuf, Mulugeta, Boobe and others I encountered along the way—and that scene in the surgical ward of St. Mary's Lacor Hospital in Gulu when the tiny child, seeing me on my knees weeping amongst the injured, toddled over and put the stumps of his arms around my neck.

Postscript

ALTHOUGH MARK TWAIN insisted that "travel is fatal to prejudice, bigotry, and narrow-mindedness," you don't have to travel physically to become informed or make a difference. *Enough Blood Shed: 101 Solutions to Violence, Terror and War* by Mary-Wynne Ashford and Guy Dauncey will inspire and alert you to positive actions on your own doorstep, where there is no shortage of need or opportunity to do something useful. If you want to travel and can afford it, consider taking not the usual Cook's Tour, where you'll be pampered and insulated, but a journey that enables you to meet ordinary people. Many organizations can offer you access to grassroots projects that will give your trip additional shape and meaning. There's no better cure for psychic numbing, or what Annie Dillard calls "compassion fatigue," than to meet those in need face to face. You'll learn to listen and come to understand the significance of this quotation, attributed to Australian Aboriginal elder Lilla Watson: "If you've come to help me, don't waste your time. If you have come because your liberation is bound up in mine, then we can work together."

A bit of exploring on the Internet will alert you to NGOs, some religious, some secular, doing excellent humanitarian work involving famine relief, poverty reduction, health support, community development, AIDS relief and prevention, grandmothers in action and programs for street children. Your contributions of time and money can save and enhance lives. L'Entraide Missionnaire is worth contacting if you want to keep abreast of events and commentary relating to sub-Saharan Africa. Simply ask Denis Tougas to put you on his e-mail list (dtougas@web.ca). Here are a few organizations, not all of them related to Africa, with worthwhile aims, good track records and a commitment to modest administrative costs:

- Academy for Peace and Development www.apd-somaliland.org
- Child Haven International www.childhaven.ca
- Dignitas International www.dignitasinternational.org
- Doctors Without Borders/Médecins Sans Frontières www.doctorswithoutborders.org
- HAVOYOCO (Horn of Africa Voluntary Youth Committee) www.havoyoco.org
- HEAL Africa www.healafrica.org
- Jesuit Refugee Service (Ethiopia) www.jrs.net
- Kiva www.kiva.org
- Les Enfants de Dieu (Kigali) www.enfantsdedieu.org.uk/donate.html
- Positively Africa www.positivelyafrica.org
- Rwanda Women's Network www.rwandawomennetwork.org
- Shalom, Educating for Peace www.shalomeducatingforpeace.org
- Stephen Lewis Foundation: Various programs, including

Grandmothers to Grandmothers, working to help the heroic women left to care for the offspring of those lost to the AIDS epidemic. www.stephenlewisfoundation.org/grandmothers_gathering_story.htmm

- Village of Hope www.hoperwanda.org
- WE-ACTX (Women's Equity in Access to Care and Treatment, Kigali) www.we-actx.org

If you are moved to assist any of the individuals or groups mentioned in this book and would like advice, here are my own coordinates:

Gary Geddes
81 Blue Heron Road, Box 13-3
Thetis Island, BC
V0R 2Y0
250-246-8176
gedworks@islandnet.com

Sources and Further Reading

Achebe, Chinua. *Things Fall Apart*. London: William Heinemann, 1958.

Aghion, Anne. *In Rwanda We Say . . . The Family That Does Not Speak Dies*. (DVD) 2004.

———. *Living Together Again in Rwanda?* (DVD) 2002.

———. *My Neighbor My Killer*. (DVD) 2009.

———. *The Notebooks of Memory*. (DVD) 2009.

Amis, Kingsley. *Lucky Jim*. New York: Doubleday, 1954.

Andrzejewski, B.W., and Sheila Andrzejewski, trans. *An Anthology of Somali Poetry*. Bloomington, IN: Indiana University Press, 1993.

Ashford, Mary-Wynne, and Guy Dauncey. *Enough Blood Shed: 101 Solutions to Violence, Terror and War*. Gabriola Island, BC: New Society Publishers, 2006.

Bailey, Stewart. "BHP, Anglo Shun Congo Risks to Expand as Copper Soars," February 7, 2006. www.bloomberg.com/apps/news?pid=newsarchive&sid=aa_GmuOoj8Ss&refer=australia.

BBC News. "Somali Poet Marches for Peace." July 21, 2003. news.bbc.co.uk/2/hi/africa/3084329.stm.

Beah, Ishmael. *A Long Way Gone: Memoirs of a Boy Soldier*. Vancouver: Douglas & McIntyre, 2007.

Berger, John. *And Our Faces, My Heart, Brief as Photos*. New York: Pantheon, 1984.

Bissett, James. "The Claims and Assertions by NATO about Kosovo Were Lies." Speech before the Canadian Hellenic Federation of Ontario, May 19–21, 2000. www.emperors-clothes.com/articles/bisset/claims.htm.

———. "History of Western Interference in the Balkans." www.deltax.net/bissett/author.htm (site discontinued).

———. "Humanitarian Intervention and the Sovereignty of a State in the New World Order: Undermined Authority and Undefined Rules of Engagement." In *The New World Order: Corporate Agenda and Parallel Reality*, edited by Gordana Yovanovich. Montreal and Kingston: McGill-Queen's University Press, 2003.

Blankenship, Steve. "The United States Intervention in Somalia: A Review Essay of *Black Hawk Down*." Paper presented at the Southeastern Regional Seminar in African Studies, SERSAS Conference, Georgia State University, Atlanta, GA, March 22–23, 2002.

Boland, Eavan. "The Politics of Eroticism." In *Object Lessons: The Life of the Woman and the Poet in Our Time*. New York: W.W. Norton, 1995.

Bono, ed. *Vanity Fair*. A Special Issue on Africa. July 2007.

Boyd, Rosalind. "Are We at the Table? Women's Involvement in the Resolution of Violent Political Conflicts (El Salvador and Uganda)." Centre for Developing-Area Studies, McGill University, June 1994 .

———. "A Brief Overview of Contemporary Uganda." CDAS Discussion Paper No. 60, Centre for Developing-Area Studies, McGill University, May 1990.

———. "The Struggle for Democracy: Uganda's National Resistance Movement," *Canadian Dimension*, 1989, 29–35.

Bradbury, Mark. *Becoming Somaliland*. Bloomington, IN: University of Indiana Press, 2008.

Burton, Sir Richard Francis. *First Footsteps in East Africa: or, An Exploration of Harar, 1856; Memorial Editions*. London: Tylston and Edwards, 1894.

Caplan, Gerald. *The Betrayal of Africa*. Toronto: Groundwood Books, 2008.

Céline, Louis-Ferdinand. *Journey to the End of Night*. Translated by Ralph Manheim. New York: New Directions, 1983.

Clarke, Walter, and Jeffrey Herbst, eds. *Learning from Somalia: The Lessons of Armed Humanitarian Intervention*. Boulder, CO: Westview Press, 1997.

Clinton, Hillary Rodham. "What I Saw in Goma." *People*, August 11, 2010. www.people.com/people/article/0,,20299698,00.html.

Dallaire, Romeo. *They Fight Like Soldiers, They Die Like Children: The Global Quest to Eradicate the Use of Child Soldiers*. Toronto: Random House Canada, 2010.

Déme, Ousmane. "Between Hope and Scepticism." *Civil Society and the African Peer Review Mechanism*. Partnership Africa Canada, 2005.

Deneault, Alain, Delphine Abadie and William Sacher. *Noir Canada: Pillage, corruption et criminalité en Afrique*. Montreal: Les Éditions Écosociété, 2008.

de Temmerman, Els. *Aboke Girls: Children Abducted in Northern Uganda*. Kampala: Fountain, 2001.

de Waal, Frans. *The Age of Empathy: Nature's Lessons for a Kinder Society; Our Inner Ape: A Leading Primatologist Explains Why We Are Who We Are*. New York: Riverhead Books, 2005.

Dillard, Annie. *For the Time Being*. New York: Viking Press, 1999.

Donini, Antonio. "The Geopolitics of Mercy—Humanitarianism in the Age of Globalization." In *The Prevention of Humanitarian Emergencies*. Edited by R. Vayrynen and W. Nafziger. New York: UNU-WIDER, 2002.

———. "Through a Glass Darkly—Humanitarianism and Empire." In *Capitalizing on Catastrophe: Neoliberal Strategies in Disaster Reduction*, by N. Gunewardana and M. Shuller. Plymouth, U.K.: AltaMira Press, 2008.

Dorfman, Ariel. *Exorcising the Terror*. New York: Seven Stories Press, 2002.

Dowden, Richard. *Africa: Altered States, Ordinary Miracles*. London: Portobello Books, 2008.

Drohan, Madelaine. *Making a Killing: How and Why Corporations Use Armed Force to Do Business.* New York: Random House, 2003.

A Duty to Protect: Justice for Child Soldiers in DRC. AJEDI-Ka (Association des Jeunes pour le Développement Intégré Kalundu), 2005.

Eriksson Baaz, Maria, and Maria Stern. "The Complexity of Violence: A Critical Analysis of Sexual Violence in the Democratic Republic of Congo (DRC)." Uppsala, Sweden: The Nordic Africa Institute, SIDA, 2010.

Farah, Nuruddin. *From a Crooked Rib.* London: Heinemann Educational Books, 1970.

———. *Maps.* London: Picador, 1980.

———. *Sardines.* London: Allison & Busby, 1981.

———. *Sweet and Sour Milk.* London: Allison & Busby, 1979.

Farmer, Paul. *Pathologies of Power.* Berkeley, CA: University of California Press, 2003.

Foley, Conor. *The Thin Blue Line: How Humanitarianism Went to War.* London: Verso, 2008.

Freire, Paolo. *The Pedagogy of the Oppressed.* London: Continuum, 1970.

Fullerton, Frederick F. "Interview with UN Veteran Antonio Donini on Lessons Learned in Afghanistan and Elsewhere." Watson Institute for International Studies, Brown University, March 4, 2004. www.watsoninstitute.org/news_detail.cfm?id=1755.

Gaarriye (Maxamed Xaashi Dhama). "A to Z." www.poetrytranslation.org/poets/Maxamed_Xaashi_Dhamac_'Gaarriye'. [The literal translation of these poems was done by Martin Orwin and Maxamed Xasan 'Alto.' The final versions, translated by David Harsent, appeared on the Poetry Translation Centre website, cited above.]

———. "Mandela." www.poetrytranslation.org/poets Maxamed_Xaashi_Dhamac_'Gaarriye'.

Gakwandi, Arthur. *Kosiya Kifefe.* Nairobi: East African Educational Publishers, 1997.

Ghazvinian, John. *Untapped: The Scramble for Africa's Oil.* Orlando, FL: Harcourt, 2007.

Gide, André. *Travels in the Congo.* Berkeley, CA: University of California Press, 1962.

Global Witness. "Lesson UN Learned: How the UN and Member States Must Do More to End Natural Resource-Fuelled Conflicts," January 2010.

Goddard, John. "Toronto Man Returns to Ethiopia a Pop Star." *Toronto Star,* November 20, 2008. www.thestar.com/entertainment/article/539409.

Gould, Lark Ellen. "A Nation of Bards." *Saudi Aramco World* 39, no.6 (1988). www.saudiaramcoworld.com/issue/198806/a.nation.of.bards.htm.

Gourevitch, Philip. *We Wish to Inform You That Tomorrow We Will Be Killed with Our Families: Stories from Rwanda.* London: Picador, 1998.

Graham, W.S. "Implements in Their Places." *Collected Poems.* London: Faber and Faber, 1977.

Griswold, Eliza. "God's Country." *The Atlantic.* March 2008.

Guest, Robert. *The Shackled Continent: Africa's Past, Present and Future.* London: Smithsonian Books, 2004.

Gutman, Roy, and David Rieff (eds). *Crimes of War: What the Public Should Know.* London: W.W. Norton, 1999.

Hardin, Herschel. *A Nation Unaware.* Vancouver: J.J. Douglas, 1974.

Harris, Geoff. "The Case for Demilitarisation in Sub-Saharan Africa." *Achieving Security in Sub-Saharan Africa: Cost Effective Alternatives to the Military.* Edited by Geoff Harris. Institute for Security Studies, 2004.

Harrison, Paul. *The Greening of Africa: Breaking Through in the Battle for Land and Food.* London: Paladin Grafton Books, 1987.

Hochschild, Adam. *King Leopold's Ghost: A Story of Greed, Terror, and Heroism in Colonial Africa.* New York: Mariner Books, 1999.

Höhne, Markus Virgil. "From Pastoral to State Politics: Traditional Authorities in North Somalia." In *State Recognition and Democratization in Sub-Saharan Africa: A New Dawn for Traditional Authorities?* Edited by Lars Buur and Helene Kyed. Basingstoke, U.K.: Palgrave Macmillan, 2007.

———. "Newspapers in Hargeysa: Freedom of Speech in Post-Conflict Somaliland," *Afrika Spectrum* 43 (2008).

———. "Somalia: Update on the Current Situation (2006–2008)." Schweizerische Flüchtlingshilfe, December 17, 2008.

Horbath, Peter. *The River Congo: The Discovery, Exploration and Exploitation of the World's Most Dramatic River.* San Francisco: Harper & Row, 1977.

Human Rights Watch. "Eastern DR Congo: Surge in Army Atrocities." November 2, 2009.

———. www.hrw.org/en/news/2009/11/02/eastern-dr-congo-surge-army-atrocities.

International Crisis Group. "LRA: A Regional Strategy beyond Killing Kony." *Africa Report,* no. 157, April 28, 2010.

———. "The Role of the Exploitation of Natural Resources in Fuelling and Prolonging Crises in the Eastern DRC." January 10, 2009. www.international-alert.org/pdf/Natural_Resources_Jan_10.pdf.

Jackson, Judy. *The Man Who Couldn't Sleep.* JudyFilms Inc., 2006.

———. *Talk Mogadishu: Media under Fire.* JudyFilms Inc., 2003.

———. *The Toughest Job in the World.* JudyFilms Inc., 1999.

———. *The Ungrateful Dead: In Search of International Justice.* JudyFilms Inc., 2006.

Jama, Musa Jama. *A Note on My Teacher's Group.* Pisa: Ponte Invisibile Edizioni, redsea-online Publishing Group, 2003.

Jefferess, David. *Postcolonial Resistance: Culture, Liberation and Transformation.* Toronto: University of Toronto Press, 2008.

Jeyifo, Biodun, ed. *Conversations with Wole Soyinka.* Jackson, MS: University Press of Mississippi, 2001.

Jordan, Robert Paul. "Somalia's Hour of Need." *National Geographic.* June 1981.

Kanza, Thomas R. *Conflict in the Congo: The Rise and Fall of Lumumba.* London: Penguin African Library, 1972.

Kaplan, Robert D. *Surrender or Starve: The Wars behind the Famine.* Boulder, CO: Westview Press, 1988.

Kapuściński, Ryszard. *Another Day of Life.* Translated by William R. Brand and Katarzyna Mroczkowska-Brand. New York: Knopf, 1987.

———. *The Emperor.* Translated by William R. Brand and Katarzyna Mroczkowska-Brand. New York: Vintage Books, 1989.

———. *The Shadow of the Sun.* Translated by Klara Gloczewska. New York: Knopf, 2001.

———. *Travels with Herodotus.* New York: Vintage Books, 2008.

Kingsolver, Barbara. *The Poisonwood Bible*. New York: Harper Flamingo, 1998.

Klein, Naomi. *The Shock Doctrine: The Rise of Disaster Capitalism*. Toronto: Vintage Canada, 2008.

Kneen, Jamie. "The Social Licence to Mine: Passing the Test." *Mines and Communities*, November 14, 2006. www.minesandcommunities.org/article.php?a=276.

Kutesa, Pecos. *Uganda's Revolution 1979-1986: How I Saw It*. London: Fountain Books, 2006.

Laurence, Margaret. *The Prophet's Camel Bell*. Toronto: McClelland & Stewart, 1963.

———. *The Tomorrow-Tamer*. Toronto: McClelland & Stewart, 1963.

———. *A Tree for Poverty*. Nairobi: Eagle Press, 1954.

Liberman, Mark. "46 Somali Words for Camel." *Language Log*, February 15, 2004. itre.cis.upenn.edu/~myl/languagelog/archives/000457.html.

Lindqvist, Sven. "*Exterminate All the Brutes*." London: Granta, 1992.

Lonsdale, John. "How to Study Africa: From Victimhood to Agency." *Open Democracy*, August 31, 2005. www.opendemocracy.net/democracy-africa_democracy/agency_2796.jsp.

MacAndrew, Heather, and David Springbett. *The Man We Called Juan Carlos*. Bulldog Films, 2001.

Mamdani, Mahmood. "The Invention of the Indigene." *Pambazuka News*, January 13, 2011. http://pambazuka.org/en/category/features/70061.

Marnham, Patrick. *Fantastic Invasion: Dispatches from Contemporary Africa*. New York: Penguin, 1979.

Martin, Brian. "Non-Violence versus Capitalism." *Gandhi Marg*, 21, no. 3 (October–December 1999).

Maskalyk, James. *Six Months in Sudan: A Young Doctor in a War-torn Village*. New York: Random House, 2009.

McKinnon, Matthew. "Kicking Up Dust." June 30, 2005. www.cbc.ca/arts/music/knaan.html.

Mealer, Bryan. *All Things Must Fight to Live: Stories of War and Deliverance in Congo*. London: Bloomsbury, 2008.

Meredith, Martin. *The State of Africa: A History of Fifty Years of Independence*. New York: Free Press, 2006.

Minow, Martha. *Between Vengeance and Forgiveness: Facing History after Genocide and Mass Violence*. Boston, MA: Beacon Press, 1998.

Moreno-Ocampo, Luis. Report to the Members States of the ICC in Kampala, May 31, 2010. www.icc-cpi.int/iccdocs/asp_docs/RC2010/Statements/ICC-RC-statements-LuisMorenoOcampo-ENG.pdf.

Mortenson, Greg. *Three Cups of Tea: One Man's Mission to Promote Peace . . . One School at a Time*. Toronto: Penguin, 2007.

Moyo, Dambisa. *Dead Aid: Why Aid Is Not Working and How There Is a Better Way for Africa*. Vancouver: Douglas & McIntyre, 2009.

Munro-Hay, Stuart. "Something Exceptional in Ethiopia: The Royal Tombs in Axum." 1998. http://archaeology.about.com/od/ironage/ig/The-Royal-Tombs-of-Aksum/Something-Exceptional-.htm.

Museveni, Yoweri. *Sowing the Mustard Seed: The Struggle for Freedom and Democracy in Uganda*. Oxford: Macmillan Education Ltd., 1997.

Naipaul, V.S. *A Bend in the River*. London: Penguin, 1979.

Neuffer, Elizabeth. *The Key to My Neighbor's House: Seeking Justice in Bosnia and Rwanda*. New York: Picador, 2001.

Orbinski, James. *An Imperfect Offering: Humanitarian Action in the Twenty-First Century*. Toronto: Doubleday Canada, 2008.

Orwell, George. "In Defence of P.G. Wodehouse." (1945) In *Collected Essays by George Orwell*. eBooks @ Adelaide, 2007.

Paris, Erna. *The Sun Climbs Slow: Justice in the Age of Imperial America*. Toronto: Knopf Canada, 2008.

p'Bitek, Okot. *Artist the Ruler: Essays on Art, Culture and Values*. Nairobi: East African Educational Publishers, 1986.

———. *The Defence of Lawino*. Translated by Taban lo Liyong. Kampala: Fountain Publishers, 2001. (Originally published as *The Song of Lawino* in 1966.)

Peck, John E. "Re-Militarizing Africa for Corporate Profit." *Z-Magazine*. October 2000.

Perkins, John. *Confessions of an Economic Hit Man: How the U.S. Uses Globalization to Cheat Poor Countries Out of Trillions*. San Francisco: Berrett-Koehler, 2004.

Power, Samantha. *"A Problem from Hell": America and the Age of Genocide*. New York: Basic Books, 2002.

Prunier, Gérard. *The Rwanda Crisis: History of a Genocide*, 2nd ed. New York: Columbia University Press, 1997.

Razack, Sherene H. *Dark Threats & White Knights: The Somalia Affair, Peacekeeping, and the New Imperialism*. Toronto: University of Toronto Press, 2004.

Redhill, Michael. *Goodness*. Toronto: Coach House Books, 2005.

Renzetti, Elizabeth. "An Interview with Jean Vanier." *Globe and Mail*, December 3, 2010.

Riva, Silvia. *Nouvelle histoire de la littérature du Congo-Kinshasa*. Paris: L'Harmattan, 2002.

Samatar, Said. *Oral Poetry and Somali Nationalism: The Case of Sayyid Mahammad 'Abdille Hasan*. Cambridge: Cambridge, 2009.

———. "The Wedding That Earned Bethesda Its Second Innings." Personal e-mail correspondence with the author, April 24, 2005.

Schweitzer, Albert. *On the Edge of the Primeval Forest*. London: Adam & Charles Black, 1951.

Scully, James. *Line-Break: Poetry as Social Practice*. Willimantic, CT: Curbstone Press, 2005.

Siebert, Charles. *The Wauchula Woods Accord: Toward a New Understanding of Animals*. New York: Scribner, 2005.

Simmons, Anna. *Networks of Dissolution: Somalia Undone*. Boulder, CO: Westview Press, 1995.

Skidmore, L., and G.P. Adams, eds. *Recent Advances in Camelid Reproduction*. International Veterinary Information Service, 2000.

Slovic, Paul. "'If I Look at the Mass I Will Never Act': Psychic Numbing and Genocide." *Judgment and Decision Making* 2, no. 2 (April 2007). journal.sjdm.org/7303a/jdm7303a.htm.

Snow, Keith Harmon. "Covert Action in Africa: A Smoking Gun in Washington," April 16, 2001. http://allthingspass.com/journalism.php?catid=10.

"Social Protection in Africa: Where Next?" Prepared by the Centre for Social Protection, Institute of Development Studies, Overseas Development Institute, Regional Hunger and Vulnerability Program, and International Development, University of East Anglia, June 2008. www.odi.org.uk/resources/download/4884.pdf.

"Somalia: 'Once the Camels Start Dying, People Are Not Far Behind.'" *The Global Report*, no. 521, March 10-16, 2009. theglobalreport.org/index2.php?article_id=6482&cat_id=172§ion=archives.

Soyinka, Wole. *Death and the King's Horseman*. New York: W.W. Norton, 2002.

Stille, Alexander. "War of Words: Oral Poetry, Writing, and Tape Cassettes in Somalia." In *The Future of the Past*. New York: Picador, 2003.

Thomas, Dylan. Thomas to Henry Treece, 1956. In *Dylan Thomas—The Collected Letters*, edited by Paul Ferris. London: Weidenfeld & Nicholson, 2002.

Thomas, Lawrence. *André Gide: The Ethic of the Artist*. London: Secker & Warburg, 1950.

Tougas, Denis. "Canada in Africa: The Mining Superpower." *Pambazuka News*, November 20, 2008. www.pambazuka.org/en/category/features/52095.

United Nations. *Report of the Panel of Experts on the Illegal Exploitation of Natural Resources and Other Forms of Wealth of the Democratic Republic of the Congo*, April 2, 2001.

Uvin, Peter. *Aiding Violence*. Sterling, VA: Kumarian Press, 1998.

Vidal, John. "How Food and Water Are Driving a 21st-Century Land Grab." *The Observer*, March 7, 2010.

"War & Peace in the 21st Century." *Human Security Report 2005: War and Peace in the 21st Century*. Vancouver: Human Security Centre, University of British Columbia, 2005.

Webb, Phyllis. "Prison Report." *The Vision Tree: Selected Poems*. Vancouver: Talonbooks, 1982.

Wellesley, Dorothy. *Letters on Poetry from W.B. Yeats to Dorothy Wellesley*. Oxford: Oxford University Press, 1964.

West, Richard. *Brazza of the Congo: Exploration and Exploitation in French Equatorial Africa*. London: Jonathan Cape, 1972.

Williams, Terry Tempest. "Twibuke: Beauty and Healing amid the Shards of Rwanda." *Orion*, September/October 2008.

Wrong, Michaela. *In the Footsteps of Mr. Kurtz: Living on the Brink of Disaster in Mubutu's Congo*. New York: Harper Perennial, 2002.

York, Geoffrey. "Cutting through Red Tape to Strike Gold in Congo." *Globe and Mail*, April 16, 2010.

Acknowledgements

I WOULD LIKE to thank those who read this manuscript or assisted me in some way during my travels in Africa: John Gilmore, Will Goede, Rosalind Boyd, Heather MacAndrew, Bernard Taylor, Ian Smillie, Claudia Perdomo, Fiona McKay, Béatrice Le Fraper du Hellen, Victor Ochen, Philip Lancaster, Chris Knight, Keith Harrison, Bill and Cathy Gilchrist, Saje Fitzgerald, Alison Shaw, Mulugeta W. Eyesus, Jane Warren, Denis Bikseha Rwinkeshe, Harald Hinkel, Yusuf Warsame, Philip Winter, Kunal Dhar, Shawn Bradley, Jean de Dieu Basabose, William Pike, Cathy Watson, Erin Baines, Bernahu Matthews, Boobe Yusuf Duale, Sarah Petrescu, Chris Nuttall-Smith, Elias Cheboud, Sam Nkurunziza, Jonathan Manthorpe, Asende Valentin, Said Samatar, Kyle Matthews, Sunita Palekar, Ephrem Gebreleul Hailu, Célestin Kakule Kiza, Ugaaso Abuukar Boocow, Maria Vargas, Gerry Caplan, Ousmane Déme, Jerusalem Bernahu, Erna Paris, Jake Wadland, Arefaynie Fantahun, Catherine-Lune Grayson, Judy Jackson, Virginie Mumbere, Mary-Wynne Ashford, Lizzie Parsons, Jeff Speed, Kubuya Muhangi, Antonio Donini, Madelaine Drohan, Hassan Madar, Martin Orwin, Gaarriye, Silvia Riva, Hadraawi, Denis Tougas, Nicholas Brass, David Lochhead, Badru Mulumba Jr., Lukangyela Bakeni Lucien, Wolde Selassie and those whose names have been changed or omitted for reasons of security. Special thanks to my agent Don Sedgwick and to my editor Barbara Pulling, whose eagle eye and painstaking attention to detail saved me a lot of embarrassment; to my publisher and long-time friend Scott McIntyre, who kept the faith and shared the risk; and to my beloved wife, Ann Eriksson, who, in addition to reading and proofing the manuscript several times, managed to sustain her necessary fictions in the face of my questionable realities.